A BOTANIST IN BORNEO
HUGH LOW'S SARAWAK JOURNALS
1844–1846

The signing of the treaty by which the Sultan of Brunei ceded Labuan to Britain, 18 December 1846. To the right is Captain Rodney Mundy, R.N. Hugh Low, who acted as interpreter in the negotiations, is one of the Europeans seated to the left.

From R. Mundy, *Narrative of Events in Borneo and the Celebes* ...,
2 vols., London, 1848, II, opp. p. 295.

A BOTANIST IN BORNEO
HUGH LOW'S SARAWAK JOURNALS
1844–1846

Edited and introduced by R.H.W. Reece

and

with notes on Hugh Low's plant portraits
by P.J. Cribb

Natural History Publications (Borneo)
Kota Kinabalu
2002

Published by

NATURAL HISTORY PUBLICATIONS (BORNEO) SDN. BHD.
(Company No. 216807-X)
A913, 9th Floor, Wisma Merdeka,
P.O. Box 15566,
88864 Kota Kinabalu, Sabah, Malaysia.
Tel: 6088-233098 Fax: 6088-240768
e-mail: chewlun@tm.net.my
Web site: www.nhpborneo.com

First published 24th April 2002.

A Botanist in Borneo: Hugh Low's Sarawak Journals, 1844–1846
    Edited and introduced R.H.W. Reece
    and with notes on Hugh Low's plant portraits by P.J. Cribb

ISBN 983-812-065-0

Printed in Malaysia.

CONTENTS

Hugh Low in 1847 on his return to England from Sarawak. From a portrait in pastels by William Montaigne, R.A., formerly in the possession of Miss Eileen Low.

INTRODUCTION

R.H.W. Reece
Murdoch University,
Western Australia

HUGH LOW was the elder son of an enterprising Scots horticulturist of the same name who came to London in about 1823 and went to work for Mackay's nursery at Clapton, then an outlying village of the metropolis. By 1827 he was describing himself as 'Foreman and Propagator' and he appears to have taken over the concern in 1831. Young Hugh was born at Clapton on 10 May 1824 and he and his brother Stuart and sister Alexina were educated privately, reflecting their father's commercial success. The Clapton nursery had good relations with the Royal Botanic Gardens at Kew and the earliest surviving letters of Hugh Low are to its then Director, Sir William Hooker, in October 1842 and August 1843, thanking him on his father's behalf for seeds and cuttings. By this time the technology of glasshouse construction (to which Clapton contributed), together with the development of the Wardian Case (a portable miniature greenhouse for collecting), meant that the cultivation of orchids was more practicable than ever before, and the high prices which could be obtained acted as a strong incentive to introduce new species from other parts of the world, especially the tropics. It was the beginning of the great period of exotic horticulture in Britain when big nurseries such as Veitch and Low vied with each other to supply wealthy fanciers and despatched professional collectors to most parts of the globe.

While there is no evidence that young Hugh had any formal botanical education, he was well-versed in botanical knowledge and the practical

1

work of the nursery where he spent two years as an assistant. It is not surprising that his father should have decided to send the nineteen year old on an expedition to a part of the world as yet unknown to collectors. On 27 June 1844 he wrote to Hooker:

> *I take the liberty of apprising you that I have determined on sending out my eldest [sic] son Hugh to collect plants seeds etc. in the various islands of the Indian Archipelago and in conformity to that resolution he will sail for Singapore on the 10th of the next month. I purpose that this should be his headquarters and that he should make excursions from thence to the adjacent islands[,] confining much of his attention to Borneo... .*

Stuart Low had earlier joined the merchant navy and it may be that this was partly for the purpose of collecting in China, where he made his first voyage. Why Hugh senior was so interested in Borneo is not clear, although he may well have read that James Brooke had recently established himself there and concluded that this would provide a good opportunity to explore its botanical resources. There appears to have been no communication with Brooke before young Hugh's departure although Hugh senior obviously had connections in Singapore where the Rajah was well-known. According, to one account, Hugh joined the East India Company's service in 'about 1840' but on meeting James Brooke on the ship out, was persuaded by him to come to Sarawak as his secretary and companion. There is no evidence that he joined the Company and the story probably originated in confusion with his brother Stuart's career. Nevertheless, it would be interesting to know how Hugh junior obtained his prior knowledge of Brooke.

Low's journal records that he sailed on the *Chieftain* from Gravesend on 17 July 1844 and did not make land until late November when he went ashore at a small island off the south-west coast of Johore to collect orchids. He disembarked at Singapore on 25 November with letters of introduction to the Governor, Colonel William John Butterworth, and other dignitaries. The following weeks he spent in a frenetic round of social visits and botanical and hunting forays into the island's plantations and jungles. His instant popularity with the small European population can be gathered from the fact that he received no less than five invitations for one evening. On 5 December, the Rajah's trading schooner, *Julia*, arrived from Sarawak with a cargo of antimony ore and Low met its captain, William Bloomfield Douglas, recording that he was 'quite delighted with his behaviour'. Four days later he left Singapore in a small cutter with Douglas to explore the Riau islands to the south, making his first acquaintance with Malay society and the Dutch colonial administration.

On his return, he prepared four large cases of *Nepenthes* or pitcher plants, together with nutmegs and mangosteens, for despatch to Clapton. Amongst these was what he believed to be a new species of pitcher plant which he described in his journal: 'The cups are small and green but most delicately and beautifully formed; this will be an acquisition.' When Stuart arrived unexpectedly on the *Greyhound* on 18 December for a short visit, the two brothers went out together and found yet another new species. Hugh also collected birdskins, butterflies, and other specimens for despatch to the British Museum, and made another despatch of six cases of plants to Clapton. On 6 January, Hugh embarked on the *Julia* with Douglas, reaching Kuching ten days later. Fascinated by the plant life observed along the Sarawak River, he wrote: 'Altogether I think appearances are very promising and I hope to find a rich harvest.' After some time on board, during which he dined with the Rajah and other Europeans every evening, he moved into the house which had originally been built for Brooke by Rajah Muda Hassim, the Brunei heir apparent from whom he had acquired the government of Sarawak in September 1841.

Low's visit to Borneo had been essentially commercial in intention and he was not employed by the Rajah during his two-year stay. Nevertheless, he became intimately involved with Brooke and his officers, and enjoyed the Rajah's patronage as Alfred Russel Wallace was to do a few years later. Apart from Brooke's great charm and his wide-ranging scientific and literary interests, Low warmed to the older man's religious views which resembled his own Unitarian leanings. Perhaps because of its isolation, Sarawak was a great talking-shop for the small community of Europeans collected there and Low was an agreeable companion. Captain Rodney Mundy, R.N., who visited Kuching in 1846, found him 'an unassuming, intelligent young man'.

One of the main topics of conversation during those long nights at the

Rajah's house was the theory and practice of governing native peoples. James Brooke's enlightened views in this area were forcefully advanced and exerted a profound influence on the young botanist whose earlier attitude towards the non-European inhabitants of Singapore had been conventionally colonial. Although his principal concerns at this point were still those of a commercial collector, Low was imbibing a store of political wisdom which he would put to good use in his subsequent career as a colonial official. What particularly impressed him was the Rajah's custom of receiving Malay and Dayak visitors in his house every evening after dinner. 'By his engaging manners and pleasing discourse', Low observed sagely, '[Mr Brooke] cannot fail to render himself more beloved than he would be supposing they were debarred from his presence.' Amongst the visitors was a deputation of people from the Sekrang River who had been devastated by the attacks made on them by Captain Henry Keppel, R.N., in August 1844. Low was favourably impressed with the Dayaks and their adaptability, and paid an early visit to a nearby village in the company of the Rajah's cousin, Arthur Crookshank. At the end of January he made his first expedition into the interior with the Rajah and Crookshank. Travelling by boat up the left hand branch of the Sarawak River, the party then walked to the Nawang Valley and visited the Singhi Bidayuh (Land Dayaks) at their mountain fastness before returning to Kuching.

Continuing to collect orchids, Low refused an invitation from Captain Drinkwater Bethune, R.N., to visit the island of Labuan (which Brooke had recommended as a British settlement), because of the need for his plants to arrive in England in the summer. Within a few weeks he was able to send off four large cases of orchids and a glass case of other species. Some of these had been obtained from the Suntah valley, fourteen miles up the left hand branch of the Sarawak River, where the Rajah had built a modest retreat for himself. Captivated by the 'delightful solitude' of unspoilt nature, Low wrote that 'nothing would be more delightful than to live and die a hermit there'. He was determined to return.

Some of the excitement experienced by Low in his early collecting in Borneo can be gained from his journal and from a letter of 12 January 1846 to Dr John Lindley describing *Vanda lowii* (now *Dymorphorchis lowii*), one of the most magnificent orchids that he found in Sarawak:

At the time I formerly sent it to you I remember having said that I expected something very magnificent in its flower, and sure I am that when it produces its spikes of flowers in England it will be the admiration of all cultivators, probably beyond any Orchid that has ever yet appeared. As I saw it nothing could exceed it in beauty; about 200 of its branches were hanging horizontally from the main stem of a large tree, from each of which depended two, three or four chains of flowers, each 10 feet in length, and sometimes 12 feet. The individual flowers are 3 inches in diameter... .

Although his responses to new plants were primarily scientific and aesthetic, it is also clear that, as the son of a professional nurseryman, he was acutely aware of their commercial possibilities. Nor were his expectations disappointed: a *Vanda lowii* sold at auction in London in 1847 for £30, a very high price indeed, and the whole consignment of Bornean orchids brought almost £400. *The Botanical Register* for 1846 also recorded the successful propagation of *Hoya imperialis* and its commercial availability at Clapton.

Apart from his botanizing and anthropologizing, Low was becoming increasingly involved in James Brooke's efforts to consolidate his fledgling state. On 16 April 1845, he left Kuching on a seventy-foot *bankong* or warboat with the Rajah's Eurasian interpreter and emissary, Thomas Williamson, on a diplomatic mission to Datu Patinggi Abdul Rahman, the Sultan of Brunei's governor of the sago-producing coastal area which provided much of the Sultan's revenue. Williamson's mission was to deliver a letter to the Patinggi directing him to resist tax demands made by the Brunei *pengiran* in the name of the Sultan. After a call on the *Tuan Melanau*, who governed the coast under the Patinggi, and an encounter with Pengiran Illudin and Nakhoda Seraddin of Brunei, the party went on up the Rejang to Sarikei where they were given an enthusiastic reception. All this provided Low with plentiful opportunities to indulge his anthropological interests. One of his more interesting passages relates to the Melanau women, whose beauty was prized far beyond Borneo. Low's conventional European notions of what constituted female beauty had been undergoing a subtle process of acclimatization:

When I first came here I used to look upon the native women with disgust; now I can easily discriminate the degrees of beauty as one resident in a European country would there. Instead of saying degrees of beauty I ought perhaps to have said plainness or rather ugliness for certainly they are not a comely race, but as I said before my ideas from constantly seeing them have become so vitiated that what we call a pretty woman we look upon with as much pleasure or nearly so as we used at the divine forms at home.

It is clear from some references in the earlier part of his journal that Low had left his heart in England, the young woman's name only being revealed as 'A D.' He was very disappointed not to have received any letter from her after arriving in Singapore.'Have been thinking lately a good deal about A D', he wrote, 'and I wish very much to hear from her'. When a letter did arrive in March 1845 it brought news that she was ill and he was clearly very concerned. He seems to have thought of her less as he became engrossed in the great adventure of Borneo because she is never mentioned again in his journal. However, as we shall see, this was not the end of the story.

In early May 1846, Low and Williamson accompanied Brooke and Captain Bethune on H.M.S. *Phlegethon* to Brunei for two weeks to investigate the coal deposits there and it was during this time that Low gained his first knowledge of the Brunei court, which was to stand him in good stead in later years. They then went to the nearby island of Labuan to see if it was suitable for settlement. Low was favourably impressed with Labuan, where he was later to spend almost thirty years of his life, and, like Brooke, imagined it as becoming another Singapore. It was on this visit that he had his first sight of Mt Kinabalu, the highest peak in South East Asia and the subsequent scene of some of his most spectacular botanical discoveries. Sir Edward Belcher, at that time surveying the Labuan coast in H.M.S. *Samarang*, was himself an amateur naturalist and may well have encouraged Low to make an expedition up the mountain.

Now closely involved through his friendship with Williamson in state affairs, Low accompanied him in October on a military expedition against the Dayaks of the Sekrang, who were said to be planning raids under the direction of two *sharif*, the part-Arab, part-Malay entrepreneurs who had assumed positions of power under the lax rule of Brunei. They were under orders 'to kill the Sereibs and disperse the boats' and Low seems to have been no mere spectator. However, the fortuitous deaths of two important Dayak chiefs, Apa Biagi and Gila Berani, and the support of another, Gasing, meant that there was no need for blood-letting after all.

Low made his two major collecting expeditions up the right and left hand branches of the Sarawak River in November and December 1845. As well as discovering a number of plants, including, some magnificent specimens of *Rhododendron brookeanum* which he named after the Rajah, he spent some time observing the customs of the Sauh and other Bidayuh tribes and it was then that he recorded his most detailed ethnographic descriptions. Obviously, he would have liked more time in the interior and received pressing invitations from various *tuai rumah* (head men) to stay, but his collection of plants still had first priority and he hurried back to Kuching to prepare them for despatch in glass cases to Clapton.

The penultimate entry in Low's journal is for 30 January 1846, recording the drowning of Williamson, who had fallen overboard from his canoe when returning from dinner at the Rajah's house the previous evening. Low's moving tribute to him is a useful reminder of the importance of this man, whose linguistic and diplomatic skills served Brooke so well that Henry Keppel referred to him as 'an excellent Prime Minister':

> *This death has visited in an awful guise our small society and taken from it by far the most amiable (Mr Brooke excepted) person of those composing it. From the beginning he has been Mr Brooke's chief assistant and the esteem he has gained for himself in the discharge of his many duties is acknowledged no less by the natives than by his superior and his friends. We all loved him and each lament him, indeed of all here he is probably the only one whom I had considered more than an acquaintance.*

The final entry in the journal (28 March) reported developments which were a serious setback for Brooke: the assassination of Brooke's principal ally in the Brunei court, Pengiran Bedruddin, together with Rajah Muda Hassim and the rest of the faction sympathetic to Brooke and the British. This marked the end of any manipulation of Brunei by Brooke and the beginning of a period of conflict between Sarawak and the sultanate, in which Low's own interests eventually diverged from those of the Brookes.

Low was an observer during Admiral Sir Thomas Cochrane's attack on Brunei in July 1846 and at the subsequent signing of a treaty with Sultan Omar Ali Saifuddin II in December which confirmed the cession of Labuan to the British Crown. Further, he was present at Labuan on 24 December to witness its official possession at the hands of Captain Mundy. Low by this time was fluent in the Malay language and Brooke had used him at Brunei as interpreter and diplomatic assistant in Williamson's place. To this extent, then, he acted as the Rajah's 'secretary'.

In June 1847 Low sailed with the Rajah for England. On the way, Brooke received news of his appointment as Governor of Labuan and Consul-General for Borneo and he evidently promised Low the position of Colonial Secretary in the new colony's administration. Brooke had by this time become a Byronic hero in England, thanks to the earlier publication of his carefully edited journals by Keppel which had already run to three editions, and Low no doubt enjoyed some of the glory reflected by his patron who was feted on all sides. While the Rajah had his portrait painted by Sir Francis Grant of the Royal Academy (which did much to enhance his dashing image), Low himself was sketched by William Montaigne, who had been a contemporary of Millais at the Royal Academy. This reveals him as an attractive and sensitive young man of twenty-three with finely-chiselled features crowned by a mass of dark, curly hair. He could easily have been mistaken for a Russian poet or a Polish composer.

From the time of his departure from England until early 1846, Low kept a detailed journal of his experiences and observations, and it was this, together with the knowledge gleaned from Brooke and Williamson in particular, which provided most of the material for his book, *Sarawak: Its Inhabitants and Productions ...*, which was published by Richard Bentley in London in January 1848, just after his departure for Singapore. Indeed, the last three chapters are based on the latter part of his journal relating to his visits to Sarikei and the upper reaches of the Sarawak River. Low wrote in his preface in December 1847 that on his arrival in England three months earlier, he had had no intention of writing the book.

He emphasised the 'circumstances of haste under which it has been written'. Apart from seeing his family, he also made a visit to Leiden in Holland to examine scientific collections there. Altogether, it was a remarkable achievement under these conditions to produce a work of such diversity and erudition. For a young man whose formal education had been somewhat limited, he also wrote with some *panache*. Exactly why he was prompted to this extraordinary effort is not clear, although it may well be that he received encouragement from Dr John Lindley, Professor of Botany at University College, London. The book is dedicated to James Brooke and although there is no indication that the Rajah bore any responsibility for it, there is no doubt that its favourable account of Sarawak's human and natural resources would have been seen by him as a timely advertisement for much-needed British investment.

A long and favourable review of the book appeared in the January 1848 issue of *Gardeners' Chronicle*, probably by Lindley who was then its co-editor, and it was also mentioned by the President of the Royal Geographical Society of London in his May address. Altogether, 1848 was a bumper year for books on Borneo. Following Low's book came Frank Marryat's *Borneo and the Indian Archipelago*, Captain Mundy's *Narrative of Events in Borneo and Celebes* (which included sections of Brooke's journal omitted by Keppel), Captain Sir Edward Belcher's *Narrative of the Voyage of H.M.S. Samarang*, and Frederick Forbes' *Five Years in China*. All of these contained extensive references to Sarawak and Brunei and Belcher appended a long account of his Assistant Surgeon's naturalizing. It may be, then, that Low's book did not make the impact that it rnight otherwise have done. It was not reprinted, although the botanical sections were extracted by Hooker and published separately the same year. Apart from its account of Bornean flora, *Sarawak* was also scientifically significant in that it constituted valid publication under the rules of botanical nomenclature for at least two new species discovered by Low.

Sarawak was the first comprehensive and authoritative description of northern Borneo and served as a standard general reference for half a century. It was also the pioneer of an entire genre of writing about the indigenous peoples of Borneo, whose custom of head-hunting has excited the curiosity of European readers to this day. In subsequent years, books by Frederick Boyle, Noel Denison, and Carl Bock about life with the

Dayaks deliberately appealed to this somewhat voyeuristic appetite but Low maintained the sober tone of a scientific observer disinclined to exploit the exotic and the sensational. In addition to his other talents, Low was a competent artist and the lithograph entitled 'Mr Brooke's Bungalow at Sarawak', which formed the frontispiece of *Sarawak*, was based on a drawing by him. He is also likely to have been responsible for the ethnographic illustrations which appear towards the end of the book. Some of the botanical paintings which he sent at different times to Lindley ended up at Kew and can still be seen there.

Low had taken back with him to England a large collection of orchids and other plants from Borneo and no doubt it was due to his careful supervision on the voyage that some were successfully propagated, most notably *Rhododendron brookeanum*, *Nepenthes ×hookeriana*, *Clerodendron bethuneanum*, and the orchid *Cypripedium lowii* (now *Paphiopedilum lowii*). However, his new appointment at Labuan brought to an end his career as a collector for the Clapton nursery and no doubt it was a great disappointment to his father that his elder son also declined to accept the responsibility of taking over the business. It is not clear if Hugh subsequently retained any financial interest in it, although this seems unlikely. On the father's death in 1863, the nursery passed to Stuart.

Colonel William Napier, better known as 'Royal Billy' due to certain pompous mannerisms, had been appointed Lieutenant-Governor of Labuan at Brooke's suggestion. A wealthy land agent and founder of the *Singapore Free Press*, Napier was a popular figure. Before his marriage, he had a daughter by a Malacca woman of Eurasian origin, and thus it was that Low made the acquaintance of young Catherine Napier on board the 44-gun frigate H.M.S. *Maeander*, which Captain Keppel had been ordered to sail to Borneo for the express purpose of delivering the new government establishment to Labuan. Also on board were James Brooke's younger nephew, Charles Johnson (later Brooke), his new private secretary, Spenser St John, and Charles Grant, who were all going out to Sarawak for the first time. On 5 May, Keppel recorded in his journal that there had been champagne and dancing to celebrate Catherine's nineteenth birthday and that there was 'something in the wind between her and Low!' By the time they reached Singapore some days later, they had decided to marry, and Keppel attended the 'cheery' wedding at St Andrew's Church on 12 August. Ten days later there was a grand ceremony to invest Brooke with the Order of the Garter and it fell to Low to read the warrant of authorization issued by Prince Albert. In the Governor's absence, Napier himself officiated as the Crown's representative and it was only Brooke's speech which prevented the proceedings from degenerating into pure burlesque after Napier had mistakenly occupied the throne symbolically reserved for the Crown.

On 29 August 1848, Brooke and the Lows embarked at Singapore on the *Maeander* for Kuching and Labuan, where they were to start their new life. This began inauspiciously with an epidemic of 'Labuan fever' (probably malaria) which killed eleven of Keppel's marines and had devastated most of the Europeans in the colony by November. Brooke himself was quite delirious at times and Keppel feared for his life. 'Fever has struck us,' the Rajah wrote to a friend, 'the greater number are miserable weak shadows, and the worst of it is, that no sooner does one recover, than another is attacked, and so the wheel of anxiety and watching continually revolves.' Only Spenser St John escaped the blight and acted as a general nurse.

From his *attap*-roofed wooden hut on the swampy shoreline, Low must have had plenty of time to reflect on his decision to abandon professional collecting for an official career. However, he was probably responsible for the relocation of the settlement on higher ground and this, together with the establishment under Governor John Scott of a garrison and permanent government buildings, did much to set the tiny colony on an even footing. In May 1849 Catherine gave birth to a son, Hugh (Hugo) Brooke Low, who was baptized by the Rajah, and a daughter, Catherine (Kitty) Elizabeth, in October 1850. Six months later the mother died of fever. There is very little information about this tragic period of Low's life, but in a story written by Sir Hugh Clifford in 1901 ('A Tale of Old Labuan'), which he said was based on fact, there is a chilling description of Low digging graves one night for his wife and fourteen other fever victims. Catherine's body was buried in the garden of the new Residency which Low had designed, possibly out of fear that Dayaks from the mainland would be tempted to take her head as they had earlier done at the island's Christian cemetery. Young Hugo and Kitty were sent off to Clapton to be brought up by their grandfather and uncle and did not see their father again for some years.

The history of Labuan's government from 1848 is a tragi-comedy of gross incompetence and bitter feuding amongst its handful of European inhabitants—a pattern best explained as a reaction to the stifling monotony and oppressive isolation of the fever-ridden island. James Brooke took little interest in his responsibilities as Governor and spent very little time there before he was relieved of the post and its useful salary in 1854. Napier, who had made a fortune as a land agent in Singapore, was an unfortunate choice as Lieutenant-Governor and exercised a disastrous land policy which ensured that Labuan did not attract any further population. He quarrelled with all his European officers and it was not long before he was not on talking terms with his son-in-law. In early 1850 he was dismissed by Brooke in connection with a young Eurasian from Singapore whom he had allowed to run the island's only 'dram-shop' (licensed premises) while occupying a government post.

Low's official duties at Labuan were not limited to those of Colonial Secretary. He also acted as Treasurer and Police Magistrate and during a number of long delays between governors, Officer Administering the Government. However, his meagre salary and his extraordinary personal generosity meant that he was always plagued by financial problems. His closest associates appear to have been Dr John Treacher, the colony's kindly but incompetent medical officer who had been his neighbour in Kuching and shared his interest in ornithology, and Lieut. J.F.A. McNair, with whom he worked on an important shell collection later given to the British Museum. He also collected more than 300 species of butterfly, twenty-nine of them new ones, which caused a great deal of interest in Britain.

It was during the governorship of George Edwardes at Labuan that Low had his falling-out with the Brookes. In July 1860 he accompanied Edwardes to Mukah in an attempt to prevent the Rajah and his nephews from wresting control of the sago-producing coastal area from Brunei, whose interests Edwardes had been instructed by London to safeguard. In his racy account of the attack on Mukah in *Ten Years in Sarawak*, Charles Brooke did not name Low as Edwardes' emissary, but as Robert Pringle has noted, there can be no doubt that he was the 'gentleman attached to the Governor's suite who brought a polite message to say, than [sic] no more fighting would be permitted on either side'. As part of his campaign to discredit the Brookes, Edwardes also forwarded to the Colonial Office Low's translation of a statement made in Brunei earlier that year by Sharif Masahor, which proclaimed his innocence of any connection with the killing of two Brooke officers, Charles Fox and Henry Steele, at Kanowit Fort in 1859. Masahor, whom Charles held responsible not only for the Mukah chief's refusal to allow in Kuching traders but for a carefully-laid conspiracy to bring down the Brooke *raj*, had barely escaped with his life in February when the Tuan Muda (as Charles was then called) sank his boat with cannon fire on the Sadong River. It is difficult to say whether Low was simply carrying out Edwardes' instructions or whether he had some sympathy for Masahor and for Brunei's continuing control of Mukah. Labuan was essentially in competition with Sarawak for the benefits of trade with Brunei, and while he had no very high opinion of the Brunei court, Low must have been acutely aware of Labuan's interests. However, unlike Spenser St John, whose early career so closely paralleled his own, he did not leave his own version of these events.

Although Edwardes' intervention temporarily thwarted the Rajah's plans and no doubt gave encouragement to Masahor, the Colonial Office recalled him for his temerity and in August 1861 Brooke was able to obtain the cession of the entire sago-producing area from Sultan Abdul Mumin. Low visited the Rajah at 'Burrator', his retirement house in Devon, in February 1862 during a year's sick leave from Labuan and some form of reconciliation was reached between the two men. The Rajah and St John certainly wrote to the Rajah's elder nephew, Captain John Brooke Johnson (better known as Brooke Brooke), in Sarawak urging him to be friendly towards Low in spite of Mukah. However, the elder nephew did not respond favourably and Charles wrote to him from Malta in March that he hoped Low had departed from England by the time he arrived. Both identified Low as being pro-Brunei. Nor can their attitude to him have been improved by the Rajah's broad hint later that year that he might install either Low or St John as a step towards the governorship of a Crown colony of Sarawak. When it transpired that the British Government would do no more than appoint a Consul and was not interested in taking over Sarawak, the idea faded, but Charles Brooke in particular seems to have had nothing further to do with Low.

Nevertheless, subsequent official Brooke references to him, notably in Bampfylde and Gould's commissioned *History of Sarawak Under Its Two White Rajahs*, acknowledged him as an authority on Borneo.

After Catherine's death, Low had formed a permanent but necessarily unofficial relationship with a Malay woman by whom he had a daughter. He may have first met her in Kuching, where a number of European officers, including St John and Treacher, had native mistresses or 'keeps'. Indeed, Dayang Kamariah, St John's mistress, was her sister. As she was invariably referred to as 'Nona Tuan Low' or 'Nona Dayang Loya' ('Nona' being the customary title in the Straits Settlements for a woman closely associated with a European), there is no indication of her Malay name, but the title *Dayang* indicates that she came from a *perabangan* or upper-class family linked with Brunei. She was a celebrated singer of *pantun* (Malay verse), which was one of the specialities of the daughters of the Kuching *datu*. Another sister was married to a prominent European merchant in Singapore.

When sixteen years old Kitty Low arrived at Labuan in December 1866, fresh from her Swiss finishing school, her father was administering the government pending the arrival of John Pope-Hennessy, a 'penniless, eloquent and horribly troublesome member for some remote Irish constituency', who had been appointed Governor. One of the first beneficiaries of Prime Minister Disraeli's more liberal attitude towards Catholics, he had previously represented West Cork at Westminster. Kitty took up residence in Low's own house but in the meantime Low had been allowing Nona Dayang Loya to visit him at the Residency, something which was regarded as scandalously immoral by those few members of Labuan's tiny European male population who had not themselves taken native mistresses. One such person was a Mr Morel, manager of the China Steamship and Labuan Coal Company's mines. He had not been on speaking terms with Low since an incident which was later reported to the Colonial Office by Pope-Hennessy himself:

> *Mr Morel said he came one day with his wife to pay his respects to the Administrator, and finding the Malay mistress of the gentleman in the house, he thought it an act of such grave indelicacy to Mrs Morel that he broke off all acquaintance wiith the temporary head of government.*

The 'grave indelicacy' was the implication that a native Malay woman, and a mistress at that, could be seen to occupy the same level in society as a middle-class European woman of married respectability.

As if Low had not been sufficiently injured by the appointment over him of a pompous little Irishman without the remotest knowledge of Eastern affairs, fate ordained that the new Governor would fall in love with and marry the beautiful but bored Kitty within a few months of his arrival. It was not long before the Governor and his father-in-law were not on speaking terms. Pope-Hennessy refused to increase his salary and Low published in a Singapore newspaper a trenchant criticism of the Governor's tax policies which were designed to make the government self-supporting. In his correspondence with the Colonial Office, whose volume was in absurd disproportion to the importance of the tiny colony, Pope-Hennessy made great play of Low's illicit relationship, citing the experience of the new Colonial Secretary's wife, Mrs Frances Slade, who 'was unable to walk up to the Coal Point Road because she found there were high officials who did not scruple to salute her when they were in the company of their Malay mistresses'. The house where Low maintained Nona Dayang Loya, her mother, and their daughter, for what he properly described as reasons of 'honour and justice and duty', was situated at Sagumau on the Coal Point Road. Pope-Hennessy himself had two illegitimate daughters in England by this time, whom he supported financially.

One evening in September 1870 when he was out riding with Kitty, Low received a message that his other daughter was ill. He immediately rode to the Sagumau house and spoke to the girl on the verandah. Kitty did not dismount but her horse was within the house compound. This grave breach of protocol was immediately reported to the Governor by the Colonial Apothecary, James McClosky, who happened to be passing by at the time. In his characteristically exaggerated style, Pope-Hennessy informed the Colonial Office that the incident had given rise to a 'grave public scandal' and described Low as being 'totally blind to the inconvenience of the public scandal attaching to his conduct'. Even the Colonial Chaplain, the Revd W.D. Beard, who detested Pope-Hennessy and believed that he was doing everything in his power to destroy Low, found the latter's action reprehensible and extracted a written promise from him that nothing like it would ever happen again.

Pope-Hennessy attempted to dispose of Low by linking him with the accusation that Nona Dayang Loya's house was being used for gambling and other illicit purposes, but a patient police watch on the house could only report that it was the scene of evening merriment, including the singing of *pantun* to musical accompaniment. In the following year the Governor finally managed to have Low indicted on a number of charges, including the accusation that he owned a house inhabited by persons 'known to the police'. A continued watch had eventually resulted in the arrest of a man for illegally selling no less than five cents' worth of tobacco outside or within Nona Dayang Loya's compound. Low was suspended from office and complained bitterly to the Colonial Office of 'perjury and tyranny', accusing Pope-Hennessy of conspiracy.

In the Pope-Hennessy papers at Rhodes House, Oxford, are the notes made by James Pope-Hennessy about his grandfather's time as Governor of Labuan and his energetic efforts to have Low dismissed from his official positions there as Colonial Secretary, Police Magistrate, etc. His principal source was the report made to the Colonial Office by the next governor, Henry Bulwer, on the charges made by his grandfather against Low. It is clear that personal animosity was behind these.

Responding to Pope-Hennessy's charge that Dayang Kamariah, who was then living with her sister at the Sagumau house after St John's departure to take up his position as British Consul in Brunei, was a common prostitute, Bulwer described her as a good and virtuous woman. He also went to some trouble to explain to the Colonial Office that *nona* was not a derogatory expression, but citing none other than Low as his authority:

> *The term 'Nona' it is said by Mr. Low is offensively translated into English as 'concubine' and I am assured by those conversant with the Malay language that the term has been abused in the translation. If it were intended to translate the word 'concubine' into Malay the term would have been 'purumpuan' or 'gundik' whereas the term 'Nona' while it signifies an unmarried woman is applied by the Malays as a term of title (and not disrespect) to women kept as mistresses by Europeans.*

Nor did Bulwer believe that Low's relationship with Nona Dayang Loya had prejudiced his official responsibilities in any way: 'while his connection with the woman Dayang Loya was well known throughout the Island, for in this small community it is impossible that such a connection, with whatever secrecy and precaution it be maintained, could be otherwise than well known. I cannot learn that he ever obtruded it upon the public observation or allowed it [at] any time to interfere with the proper discharge of his magisterial duties.'

Bulwer exonerated Low on all of the charges, evidently accepting that Pope-Hennessy's intention had also been to implicate him with the allegedly illegal actions of Nona Dayang Loya's family who lived with her in the house that he obtained for her at Sagumau. The petty details of all this are not worth repeating but they reveal something of how relationships between European men and local women were conducted at the time.

After St John left Brunei in 1863 for his new post as British *chargé d'affaires* in Haiti. he continued to support Dayang Kamariah and their surviving son Sulong (two other children had died during a fever epidemic in Labuan), whom he sent to Britain for training as a civil engineer in 1866 and then helped obtain a job somewhere in the Malay States. Kamariah herself went to Singapore and lived with another European there before dying in Labuan in about 1872. Spenser St John's brother James, who was a surveyor in Labuan, also had a family by a Malay woman there.

St John never revealed how it was that James Brooke had forced him out of Sarawak in 1855 but it probably had something to do with his open relationsip with Dayang Kamariah and the fierce hostility this aroused from the first Bishop of Labuan and Borneo, F.T. McDougall, and his wife Harriette.

There had already been an embarrassing family connection with the Borneo Church Mission. McDougall unwisely attempted to convert Kamariah's younger brother, 'John', to Christianity and actually despatched him to London with his assistant, Walter Chambers, in 1858 to get him away from the influence of his Muslim relatives. In the archives of the Society for the Propagation of the Gospel (SPG) in Rhodes House there is a letter from McDougall to Ernest Hawkins informing him of Chambers' imminent arrival in London:

> *I have sent him with one of our boys, John, a Malay pengiran to prevent him being taken away by his friends and made a Mahometan of. May I ask your interest on his behalf—Mr Chambers will explain my views with regard to him.*
>
> *P.S. The week after Mr Chambers sailed a letter came for John from his Malay relatives which I have not opened but have every reason to believe is a command for him to go there [Labuan].*

Not surprisngly. the experiment was a failure and 'John' was sent back from London to Labuan where Low found a position for him as turnkey of the colony's gaol. When McDougall learnt of this on a subsequent visit to Labuan, he prevailed on Governor Callaghan (Bulwer's successor) to dismiss him. St John and others who were critical of McDougall for other reasons no doubt cited John's case as evidence of the Bishop's political *naïveté* in attempting to convert Muslims to Christianity. John's appointment as turnkey also presented Pope-Hennessy with further ammunition to discredit Low.

Low was eventually cleared of the charges against him and reinstated but the bitter war between the two men was only ended by Pope-Hennessy's transfer to governorship of the Bahamas in 1871. Perhaps the most spectacular episode had been the Governor's issue of a writ of *habeas corpus* against Low for illegally detaining his child by Kitty after she had returned with all her belongings to her father's house. This was in response to Pope-Hennessy's petulant refusal to allow her and Low to visit Kuching, where Hugo had just arrived to join the Rajah's service.

In spite of his range of official responsibilities and the vexing distractions described above, there was still time for Low to pursue his botanical and other scientific interests. He made a number of visits to Brunei, Lawas, and other neighbouring rivers on the mainland where he discovered *Dendrobium lowii* and various pitcher plants, including *Nepenthes bicalcarata*. And in 1851, accompanied by his Chinese servant and a Dusun guide called Lemaing, he climbed Mt Kinabalu, discovering new species of orchid and rhododendron and the giant pitcher plants unique to the area. At the bleak and rocky summit, which he could only reach barefooted, he 'finished a bottle of excellent madeira to Her Majesty's health and that of my far distant friends' and planted the upturned bottle in a conspicuous place to mark his achievement. Access to the mountain had only been possible because of Low's diplomacy with the hill Dusun, or Ida'an, who had never seen Europeans before and were distinctly unfriendly. Thomas Lobb, another botanical collector employed by Veitch and Sons, was prevented by them from making an ascent five years later. However, most of Low's carefully collected specimens were jettisoned by his Dusun porters as they made their way down the mountain in monsoon rain and there was insufficient time to make a full investigation of the botanical resources of the area.

In his account of the Kinabalu expedition which was published in *The Journal of the Eastern Archipelago* in 1852, Low made no mention of his prior encounter with a flotilla of Illanun raiders at the mouth of the Tuaran River. This is the description he gave of it almost fifty years later when he had had more time to reminisce about his life in Borneo:

We were suddenly alarmed by hearing the sound of large gongs towards the mouth. We of course expected the sound to come from approaching enemies, as at that time, 1851, friends were rarely met with in those seas. We had not long to wait. Soon a two-masted vessel, with double banks of oars, pulled round the point of land. and was quickly followed by five others, all gaily decorated with flags and streamers, and having their decks covered with armed men. We recognised them at once as Llanun pirates, and I instructed my pilot to hail them and inquire who they were and what they wanted. A very handsome young man, of about twenty-eight, in a coat of armour formed from the plates of horn of the water buffalo, connected together by brass chainwork, standing in front of his companions, answered, 'I am the Sultan Si Mirantow, of Layer-layer, and having heard that Mr. Low is in the river, I have come to pay him a friendly visit.' We were in my boat seventeen men in all, sixteen of them being Brunei Malays, and the relief of receiving this reply may be easily imagined. We immediately invited the chief to an entertainment, killed the fatted calf which had that morning been presented to us by the people of the village, and held high festival till the evening, when we parted on the best of terms with our interesting guests. I never had the opportunity of meeting this agreeable young corsair again, as he was killed shortly after in an action with a Spanish gunboat.

Low returned to the mountain with his friend Spenser St John in April 1858 but their small party was challenged by a large group of Ida'an who claimed that they had suffered bad crops ever since Low's earlier ascent and demanded that he supply a slave in compensation. St John's account of the incident makes it clear that for all his mild manners, Low possessed considerable strength of nerve:

> *As the Ida'an were shaking their spears and giving other hostile signs, we thought it time to bring the affair to a climax; so I ordered the men to load their muskets, and Mr Low, stepping up to the chief with his fivebarrelled pistol, told the interpreter to explain that we were peaceable travellers, most unwilling to enter into any contest; that we had obtained the permission of the Government of the country, and that we were determined to proceed; that if they carried out their threats of violence, he would shoot five with his revolver, and that I was prepared to do the same with mine; that they might, by superior numbers, overcome us at last, but in the meantime we would make a desperate fight of it.*
> *Thus closed the scene... .*

St John managed to make the summit, where he found Low's bottle undisturbed, but Low's feet caused him great trouble and he had to be carried back in a litter. Even then, he managed to hobble to St John's assistance, brandishing his revolver, when there was another incident with the Dusun. It was also revealing that, after seven years, Low recognized the voice of his original guide, Lemaing. Like James Brooke, Low had an extraordinary ability to remember names and voices and this, together with his mastery of idiomatic Malay and *jawi* (Arabic) script, helps to explain his influence.

The expedition had been a botanical disaster and in July of that year Low and St John made another ascent, using Low's original route via the Tuaran River and Kiau. This time Low was able to make a large collection of plants, including the four great *Nepenthes* which were illustrated in St John's *Life in the Forests of the Far East*, which he published at his own expense in London in 1862. St John's entertaining account of the two expeditions was also based on Low's own journals which do not seem to have survived. Always generous with information and collected material, Low subsequently gave Frederick Burbidge detailed instructions on the location of the pitcher plants on Kinabalu when he was visiting Borneo in 1877 on a collecting expedition for the Veitchian Collection at Chelsea. Thus it was Burbidge rather than Low who succeeded in introducing the giant *Nepenthes* to England.

In the more practical field of economic horticulture, Low had a number of significant successes and his garden at Labuan was famous for its pomeloes, mangoes and mangosteens. He introduced the Balinese pomelo (*limau Bali*) and it is this fruit, well known throughout Malaysia, which is proably his most widely appreciated legacy. According to James Pope-Hennessy, Low also had his own aquarium, aviary, and collections of shells, butterflies, moths, snake skins and stuffed animals, and kept pet monkeys in his house. 'He seems to have preferred the company of Malays to that of Europeans, and of his pet animals to either. He once wrote to ... Kitty that he loved only two creatures in the world—his wah-wah [gibbon] monkey, Eblis, and herself.'

In 1876, by which time Low might well have become resigned to remaining at Labuan for ever, he was appointed British Resident at Kuala Kangsar in Perak. The first Resident, James Birch, had been assassinated by Malays in November 1875 and the security of the potentially rich area was essential to British interests in the Malay States. Low's task was a difficult and challenging one, and he rose to the occasion in his own quiet and measured way. Indeed, he is best known for his achievements in Perak, where he remained until 1889. (It is unnecessary to recount his official career, which has been discussed extensively by the historians of nineteenth century Malaya.) His Perak journal for two months of 1877 has also been published and this, together with Isabella Bird's vivid description of life at the Kuala Kangsar Residency two years later, provides some useful insights into his *modus operandi*. Of him she wrote:

> *Mr. Low is only a little over fifty now, and when he first came the rajahs told him that they were 'glad the queen had sent them an old gentleman'. He is excessively cautious, and, like most people who have had dealings with Orientals, is possibly somewhat suspicious, but his caution is combined with singular kindness, of heart, and almost faulty generosity regarding his own concerns... .*

During those busy years when he worked a fourteen-hour day, Low still managed to maintain a keen interest in economic horticulture, conducting experiments with coffee, cinchona, pepper, tea, sugar, rice, and rubber. In October 1877, Henry Murton of the Singapore Botanic Gardens brought to Kuala Kangsar ten *Hevea brasiliensis* plants which had been obtained from Ceylon, and nine of these were successfully planted in the Residency garden. Low subsequently made test plantings in different parts of the state and obviously envisaged the economic importance of this new crop. Cattle-breeding was another interest and he was responsible for introducing Jerseys and Alderneys as well as Nellore cattle from India.

Before retiring to England in 1889, Low was given a mission designed to exploit his close knowledge of Borneo affairs and his legendary diplomatic skills. The British Government had finally decided to make formal treaties with Rajah Charles Brooke and the Sultan of Brunei which would regulate relations with the two Bornean states. Low had little difficulty in persuading the Sultan to sign the treaty of 'protection'. On an earlier visit he had obtained a copy of the *Silsilah Brunei*, the chronicle of descent of the rulers of Brunei, and his annotated translation of this important *jawi* manuscript had subsequently appeared in the *Journal* of the Straits Branch of the Royal Asiatic Society which he helped to found.

At the end of his time in Perak, Low evidently hoped to be appointed Governor of the Straits Settlements but his official reward was the C.M.G. bestowed on him in 1879, followed by the K.C.M.G. in 1883 and the G.C.M.G. in 1889 in recognition of his recent services in Brunei. During these years he maintained his links with the Colonial Office, which frequently consulted him on Malayan and Borneo matters. For example, he facilitated the visit to England of Sultan Idris of Perak whose succession he had earlier engineered. In 1891 he was approached by shareholders of the Peruvian Corporation for advice on the agricultural potential of land allocated to them by the Peruvian government.

Low's administrative achievements at Labuan are difficult to assess, since at no time did he exercise unfettered authority and responsibility. Although he was in charge for eight of the twenty-seven years that he spent on the island, it may have seemed pointless to him to initiate policies which a new Governor might well sweep away with his new broom. One sympathetic contemporary observed that 'Labuan never appeared to give him the scope necessary for his abilities', but it is more likely to have been a combination of official constraints and Low's own quiet and unassuming temperament. Certainly, when he was given full authority in Perak, his financial, political, and judicial achievements became almost legendary, providing a 'model of administration for the Malay States'. These are the words of his *protégé* and successor in Perak, Sir Frank Swettenham, who owed a good deal to him. Sir Richard Winstedt was more circumspect in his assessment of Low as an originator of policy, but did not question his administrative abilities and his preference for persuasion rather than intimidation in his management of the Malays. Low's gradualist approach to the abolition of debt-slavery (*orang berhutang*) in Perak is a good example of his low-key and non-confrontational methods.

In a warm testimony to his mentor shortly after his death, Swettenham wrote:

> *The real value of Sir Hugh Low's work was to be found in the influence he exerted to prove to the Malays the meaning of justice, fair dealing, and consideration of their claims and their prejudices. That influence was not less firmly and wisely used to teach his officers a lesson of strict integrity, and to insist upon their treating the natives with the same courtesy and consideration which he showed himself. Sir Hugh Low understood what others in authority should never forget, that the only way to deal with a Malay people is through their recognised chiefs and headmen. To gain their co-operation it is necessary to show them at least as much consideration as if they were Europeans, and infinitely more patience. Moreover, they should be consulted before taking action, not after.*

Between them, Low and Swettenham were the architects of the system of indirect rule in British Malaya. Low's philosophy was best expressed in a letter to Sir William Robinson, Governor of the Straits Settlements, in 1878 in which he wrote of his Perak experience: 'we must first create the Government to be advised.' There was more than an echo here of James Brooke's policy in Borneo, and it was a principle that was to make Low one of the great pro-consuls of late nineteenth century British imperialism.

In 1885 when he was sixty-one, Low married Anne Douglas, daughter of Major-General Sir Percy Douglas, fourth Baronet of Monkseaton in Northumberland and the mysterious 'A D' of his journal. Whether she had waited for him all that time is not clear (she had not married), but they seem to have found happiness together at last.

Two years later, Low received the tragic news that Hugo had died of pneumonia while on leave in London, aged thirty-eight. Hugo had failed the examination for the Indian Civil Service, but after entering the Rajah's service he became one of the most outstanding officers of the Brooke era. Like his father, Hugo won the confidence and affection of the people amongst whom he worked. However, he was also a man of action. One of his earliest postings was at Sibu Fort, where he helped to repel an attack by the Dayak leader, Lintong, and three thousand of his followers. He served for a time at Simanggang but was later posted to the Rejang River, where he became particularly interested in the Kayan and other minority groups. An enthusiastic ethnographer and a skilled linguist, he assembled an unrivalled store of information about them and his collection of artefacts formed the basis of the Sarawak Museum's ethnographical exhibits. The ethnographic data which he took back to England on what was to be his last visit was later edited by Henry Ling Roth and published in two volumes in 1896 as *The Natives of Sarawak and British North Borneo*.

During his retirement years in Kensington, Low maintained his links with Kew, particularly with Sir Joseph Hooker and the new Director, W. Thiselton Dyer. It seems likely that he had earlier supplied considerable information to Hooker for his work on economic botany, although this has never been properly acknowledged. He was elected Fellow of the Anthropological Society, the Zoological Society and the Linnean Society and served on the Council of the last-mentioned body in 1896. He assembled a unique collection of pamphlets on Borneo (subsequently acquired by Rhodes House Library, Oxford), but apart from his journals for 1844–46 and 1877 and his official correspondence with the Colonial Office from Labuan and Perak, his personal papers have disappeared. This may be one of the reasons why his identity has been confused with that of his son Hugo by such authorities as the German bibliographer of Borneo, Karl Helbig.

Although he did not produce any scientific papers based on his own discoveries in Borneo and Malaya, Low kept abreast of botanical and horticultural developments and maintained a close connection with the family nursery, which had been moved from Clapton to Bush Hill near Enfield in 1881 due to air pollution and the need for more space. Not surprisingly, orchids were a speciality of the nursery, which his brother Stuart had managed since their father's death, and this tradition remained in the family until the death in the 1960s of Stuart's grand-daughter, Eileen Low, who had a nursery at Tunbridge Wells. The name was kept alive by *Stuart Low (Enfield) Ltd.*, which continued to cultivate exotic plants and ferns until recent times.

A full account of Hugh Low's botanical and horticultural achievements is long overdue and will no doubt reveal him as one of the important nineteenth century collectors and horticulturists. Apart from the many new species of *Nepenthes* which he discovered, the most noted of which (*N. edwardsiana*, *N. rajah*, *N. villosa*, and *N. lowii*) were described by Hooker in *Icones Plantarum* (1852) and *Transactions of the Linnean Society* (1859, 1860), he was also responsible for bringing to public knowledge many new species of orchid and rhododendron. Perhaps the most famous of the orchids, some of which were named after him, are *Coelogyne pandurata*, *C. asperata*, *Dendrobium lowii*, *Paphiopedilum lowii*, and *Dimorphorchis (Vanda) lowii*. As Burbidge wrote in an appreciative obituary in *Gardeners' Chronicle*, 'to have discovered such a regal orchid as *Vanda ... lowii* and the great mountain pitcher plants of Borneo would alone suffice to give Sir Hugh Low worldwide fame as a botanist'. Collections of his botanical specimens were lodged with the Herbarium at Kew and at Cambridge, Leiden and Vienna. Between them, Hooker and Otto Stapf were responsible for the scientific identification of most of his botanical discoveries and it is worth noting that in Borneo itself, *daun lo* and *bunga lo* became the vernacular names for the many indigenous orchids that he collected there. Low's name is also commemorated in at least one species of bird and numerous insects.

As for his work in horticulture, perhaps the best-informed appreciation was given in 1911 by Henry Ridley, then Director of the Singapore Botanic Gardens, who wrote to a friend: 'Sir Hugh Low was indeed a great agriculturalist and must rank next to Raffles as the greatest man we have had here. If he had remained here, agriculture would have been a century ahead of its present day status but much of his work was destroyed after he left.'

Low died at Hanbury Gardens near Alassio on the Italian Riviera on 18 April 1905, shortly before his eighty-first birthday, and the numerous obituaries paid tribute to his botanical and administrative achievements, emphasizing the almost self-effacing modesty which had always marked his demeanour. Lacking the flamboyant style of a James Brooke or a Frank Swettenham, Low was a quiet and efficient achiever who mastered the art of what subsequently became known as 'native administration'. 'Tuan Il Low', as he was known in Borneo, was a brilliant practitioner of the principles which the White Rajah had promulgated without any sustained interest in applying. A gifted naturalist whose practical training had instilled habits of discipline and order unknown to Brooke, he eschewed the fame that botanical science could have brought him for the practical and democratic benefits of horticulture. It is poetic, then, that the highest peak of Kinabalu should bear aloft the name of this talented but unassuming man who always seemed surprised when people remarked on his achievements but never received the full recognition due to him.

There is a barely legible letter in Rhodes House Library from Anne Low to Kitty's sister-in-law, Una Pope-Hennessy, dated 27 May 1905, which gives a poignant description of Low's death:

Una Dear,

> *I got your little p.[ost] c.[ard] just after he [Hugh] left me. I asked James Campbell—is it not wonderful that he [illegible] 'chanced' to be with me who can help as no one else could have done? I asked James Campbell to wnte to Prince Kropotkin your wish. Then if you write to Prince K. he will know and understand.*

> *I loved to get your letter just now. Some words [illegible] thoughts to me and him. He was always so fond of you—you were always so sweet to him.*

> *I had been very ill. We had come here to get Hugh to go home. Then quite suddenly he left me. A glorious last day we had among the flowers he loved—a [illegible] that [illegible] he had been [illegible] before and had pulled through. I thought he would [illegible] leave me. Half an hour or less bad suffering borne as HE would bear it.*

Then he thought I had relieved him but it was a gentle death. He slipt away holding my hand. His Lucy and I did everything for him and he looked so unspeakably beautiful and dignified. The Italian Dr who came to verify the death exclaimed at his Beauty. I was proud and pleased. I always loved to have him admired.

We walked after him James and his Lucy and I up a steep little path, next day, to a quiet little corner among the roses. The gardeners and the gardens he loved carried him along and Lucy, bless her! threw one of our Malay sarongs over my old Chief. His Malays will like to hear that he went so to his rest... .

Anne wrote again to Una a few weeks later saying that she wished to give £150 or £200 to '[some] scientific object in his [Low's] name, something connected with Botany or Natural History, his only passion besides me who loved them only for his sake (and was half jealous of them sometimes)...'.

The first letter helps to solve another mystery—the identity of the daughter Low had by Nona Dayang Loya in Labuan. Anne's references to 'his Lucy', who was apparently living with them, suggests that this was the same girl whose existence had been one of the issues in Pope-Hennessy's campaign against Low in 1867. What happened to Anne and Lucy after Low's death is not known but they probably went back to live at Anne's last address, Cranborne Grange, Micheldever, Hampshire. What had become of Nona Dayang Loya when Low left Labuan for Perak is also unknown. Nor is there any indication of the fate of Low's journals, including his account of his ascent of Kinabalu, which had been the basis of much of St John's *Life in the Forests of the Far East*. Only one (1844–46) has survived in the form of a typed transcript made by James Pope-Hennessy and is now published for the first time.

The original, a 'large ledger' used by R.E. Arnold in a series of articles published in *The Orchid Review* in 1932, was described by him then as 'in parts ravaged by the effects of time, sea-water and, perhaps by the tropical climate in which Sir Hugh lived; in places it is with difficulty deciphered, in some others it is remarkably plain, and through it breathes the spirit of a man richly endowed with understanding and love of plant life...'. It was also used by Lt. Col. C.F. Cowan in his fine 1968 appreciation of Low but has since disappeared.

27

Hugh Low at Labuan in 1869.

Obituaries of Hugh Low

The Times, 22 April 1905, p. 9.
Orchid Review, June 1905, pp. 182–183.
Transactions of the Linnean Society of London, 117th session (1905), pp. 39–42.
Gardeners' Chronicle, 29 April 1905, pp. 264–265.
Agricultural Bulletin of the Straits Settlements and Federated Malay States, Vol. 4 (1905), p. 239.
Journal of Botany, British and Foreign, Vol. XLIII (1905), p. 192.
Perak Government Gazette, Vol. XVIII, No. 18 (16 June 1905).

Works by Sir Hugh Low

(i) Published works

1848. *Sarawak; Its Inhabitants and Productions: Being Notes during a Residence in that* Country with H.H. The Rajah Brooke, London, Richard Bentley; reprinted Singapore, Oxford University Press, 1988, with an introduction by R.H.W. Reece.

1848. [Botanical extracts from Sarawak], W.J. Hooker (ed.). *Companion to the Botanical Magazine* (new ser.), pp. 16, 21 and 2435, in *Supplement to Curtis's Botanical Magazine*, Vol. 74.

1852. 'Notes of an Ascent of the Mountain Kina Balow', *The Journal of the Indian Archipelago and Eastern Asia*, Vol.6, Pt. l, pp. 1–17.

1880. 'Selesilah (Book of the Descent) of the Rajahs of Brunei', *Journal of the Straits Branch of the Royal Asiatic Society*, Vol. 5, Pt. 1, pp. 1–35.

1883. Letters to Colonial Secretary, Straits Settlements, 14 January and 26 April 1882, in I. Bird, *The Golden Chersonese and the Way Thither*, London, John Murray, Appendix C, pp. 375–379.

1899. *India, Ceylon, Straits Settlements, British North Borneo, Hong Kong. British Empire Series, Vol. I*, London, Kegan Paul, Trench, Trubner & Co. Ltd., pp. 462–97.

1954. 'The Journal of Sir Hugh Low, Perak, 1877', ed. by E. Sadka, *Journal of the Malayan Branch of the Royal Asiatic Society*, Vol. 27, Pt. 4, pp. 1–104.

1965. 'The Murder of Fox and Steele: Masahor's Version', ed. by R. Pringle, *Sarawak Museum Journal*, Vol. 12, Nos. 25–6, pp. 125–7.

Reviews of *Sarawak*

Journal of the Royal Geographical Society of London, Vol. 18 (1848), p. lvi (W.J. Hamilton).
Gardeners' Chronicle, 22 January 1848, pp. 54–5 (Dr J. Lindley).

(ii) Manuscripts

Letters, 1842–92, Royal Botanic Gardens, Kew.
Journal, 1844–46, Rhodes House Library, Oxford (typescript of original which cannot be located).
Notes on location of Kinabalu Nepenthes, 30 October 1877, Trinity College Library, Dublin.
Correspondence with Colonial Office, 1848–99, Public Record Office, Kew.

References

Arnold, R.E. (1932). 'Sir Hugh Low in Sarawak', *Orchid Review*, June–October: 163–166, 208–211, 236–238, 267–270, and 292–294.

Baring-Gould, S. and Bampfylde, C.A. (1909). *A History of Sarawak Under Its Two White Rajahs.* Henry Sotheran, London.

Barr, P. (1977). *Taming the Jungle: The Men Who Made British Malaya.* Secker & Warburg, London.

Belcher, E. (1848). *Narrative of the Voyage of H.M.S. Samarang, during the Years 1843–46.* 2 vols. Reeve, Benham, London.

Bird, I.L. (1883). *The Golden Chersonese and the Way Thither.* John Murray, London; reprinted Oxford University Press, Kuala Lumpur, 1982.

Bock, C. (1881). *The Head-Hunters of Borneo…*, Sampson Low, London; reprinted Oxford University Press, Singapore 1985, with an introduction by R.H.W. Reece.

Boyle, F. (1865). *Adventures among the Dayaks of Borneo*. Hurst & Blackett, London.

Brooke, C. (1866). *Ten Years in Sarawak.* 2 vols. Tinsley, London; reprinted Kuala Lumpur, Oxford University Press, 1990, with an introduction by R.H.W. Reece.

Buckley, C.B. (1965). *An Anecdotal History of Old Times in Singapore…*, University of Malaya Press, Kuala Lumpur; reprinted Singapore, Oxford University Press, 1984 (originally published in 2 vols., Singapore, Fraser & Neave, 1902).

Burbidge, F. A. (1880). *Gardens of the Sun*. John Murray, London.

Clifford, H. (1926). *In Days that are Dead*. John Murray, London.

Coates, A.(1987). *The Commerce in Rubber: The First 250 Years*. Oxford University Press, Singapore.

Coats, A.M. (1969). *The Quest for Plants: A History of the Horticultural Explorers*. Studio Vista, London.

Cowan, C.F. (1968). 'Sir Hugh Low, G.C.M.G. (1824–1905)', *Journal of the Society for the Bibliography of Natural History* 4(7): 327–343.

Denison, N. (1879). *Jottings Made during a Tour amongst the Land Dyaks of Upper Sarawak, Borneo, during the Year 1874*. Mission Press, Sirgapore.

Desmond, R. (1977). *Dictionary of British and Irish Botanists and Horticulturists …*. Taylor Francis, London.

Forbes, F.E. (1848). *Five Years in China: From 1842 to 1847. With an Account of the* Occupation of the Islands of Labuan and Borneo by Her Majesty's Forces. London.

Gullick, J. (1964). *Malaya*. 2nd edition. Ernest Benn, London.

Gullick, J. and Hawkins, G. (1958). *Malayan Pioneers*. Eastern Universities Press, Singapore.

Hahn, E. (1953). *James Brooke of Sarawak*. Arthur Barker, London.

Hall, M. (1958). *Labuan Story…* Chung Nam Printing Co., Jesselton.

Helbig K.M. (1955). 'Die Insel Borneo in Forschung und Schrifttum', *Mittezlungen der Geographischen Gesellschaft in Hamburg*, Band 52: 105–395.

Hooker, J.D. (1852). 'New Species from the Kinabalu etc.', *Icones Plantarum* 9: 883–898.

Hooker, J.D. (1859). 'On the Origin and Development of the Pitcher of Nepenthes, with an with an Account of Some New Bornean Plants of the Genus', *Transactions of the Linnean* Society 22: 410–424.

——————. (1896). 'Illustrations of the Floras of the Malayan Archipelago and of Tropical Africa', *Transactions of the Linnean Society* 23: 55–72.

Keppel, H. (1853). *A Visit to the Indian Archipelago, in H.M. Ship Maeander, With Portions of* the Private Journal of Sir James Brooke, K.C.B. 2 vols. Richard Bentley, London.

——————. (1899). *A Sailor's Life under Four Sovereigns...*, 3 vols. Macmillan, London.

Lemmon, K. (1962). *The Covered Garden*. Museum Press, London.

[Lindley, J.]. (1846). 'New Garden Plant: Hoya imperialis', *Botanical Register* 32: 68A.

Loh, C.Y. (1970). 'Men in the Sarawak Civil Service, 1843–1941... VIII. Henry Skelton, 1866–1873 and Hugh Brooke Low, 1869–1887'. *Sarawak Gazette*, 30 November: 212.

Luping, M., Chin, W. and Dingley, E.R. (eds.). (1978). *Kinabalu— Summit of Borneo*. Sabah Society Monograph, Kota Kinabalu.

Marryat, F.S. (1848). *Borneo and the Indian Archipelago with Drawings of Costume and* Scenery. Longman, Brown, London.

Moulton, J.C. (1915). 'An Account of the Various Expeditions to Mt. Kinabalu', *Sarawak Museum Journal* 2(6): 141.

Mundy, R. (1848). *Narrative of Events in Borneo and Celebes, down to the Occupation of* Labuan: From the Journals of James Brooke Esq., 2 vols. John Murray, London.

Nelson, E.C. (1991). 'The Waxing of a Glorious Rajah', *Kew Magazine*, 8(2): 81–89.

Pringle, R. (1970). *Rajahs and Rebels: The Ibans of Sarawak under Brooke Rule, 1841–1941*. Macmillan, London.

Pope-Hennessy, J. (1964). *Verandah: Some Episodes in the Crown Colonies 1867–1889*. Allen & Unwin, London.

Reece, R.H.W. (1985). 'A "Suitable Population": Charles Brooke and Race-mixing in Sarawak', *Itinerario* 9(I): 67–112.

Reece, Bob. (1997). 'The Loves of Hugh Low'. *Borneo Research Bulletin* 28: 55–59.

Reinikka, M.A. (1972). *A History of the Orchid.* University of Miami Press, Coral Gables.

Ross, J.D. (1911). *Sixty Years: Life and Adventure in the Far East.* 2 vols. Hutchinson, London.

Roth, H.L. (1896). *The Natives of Sarawak and British North Borneo*, 2 vols. Truelove & Hanson, London.

St John, S. (1862). *Life in the Forests of the Far East*, 2 vols. Smith Elder, London; reprinted Oxford University Press, Singapore 1986, with an introduction by T.H. Harrisson.

——————. (1879). *The Life of Sir James Brooke, Rajah of Sarawak, from his personal papers* and correspondence. Blackwood, London; reprinted Oxford University Press, Kuala Lumpur 1994, with an introduction by R.H.W. Reece.

——————. (1862). 'Observations on the North-west Coast of Borneo', *Journal of the Royal Geographical Society* 32: 217–233.

Stapf, O. (1894). 'On the Flora of Mount Kinabalu, in North Borneo', *Transactions of the Linnean Society* 4 (2nd ser., Botany): 69–263, pls. 11–20.

Swettenham, F. (1907). *British Malaya...* John Lane, London.

Templer, J. (ed.). (1853). *The Private Letters of Sir James Brooke, K.C.B....*, 3 vols. Richard Bentley, London.

Tarling, N. (1971). *Britain, the Brookes and Brunei.* Oxford University Press, Kuala Lumpur.

Tarling, N. (1982). *The Burthen, the Risk, and the Glory: A Biography of Sir James Brooke Brooke.* Oxford University Press, Kuala Lumpur.

Van Steenis, C.G.G.J. (ed.). (1953). *Flora Malesiana*, Ser. I, Vol. 1, Groeningen, Noordhoff Kolff, pp. 331–332.

Winstedt, R. (1948). *Malaya and Its History.* Hutchinson, London.

Journals of Hugh Low
November 1844–March 1846

1844

Nov 19th [1844]. From this day to Nov 22nd we were found generally with light winds from [the] North. On the 20th we were at anchor at East of Tanjong Suan or Cape Rachado and on the night of the following day in again anchoring near Tanjong Boulers the most southern promontory of Asia. We weighed [anchor] on the following morning but had scarcely made sail when it again became calm and I took a Kling sampan and went on shore accompanied by the second mate to a small Islet called Copper Island by the English and Pula[u] Chalot by the natives. I saw on it many plants which I had known in England, amongst them a Hoya, Dendrobium cumminativum (? = crumentatum) and a pink flaming Dendrobium [? secundum] and the Bird's nest fern are in great luxuriance.

Nov 22 Friday. Weighed in the afternoon and anchored off St. Johns Island in sight of the shipping in Singapore Roads.

Nov 23 Saturday. A squall carried us as far as the Roads but our Captn. being ill the mate anchored about 2 miles from the other ships. In the evening the skipper was much troubled with the cramp and I sent the cutter on shore for Dr Martin who reached the ship at about ll o'clock and gave Captn. Smith some medicine from the Ship's chest. He seems a most gentlemanly man and promises me all the assistance in his power. I am to dine with him tomorrow.

Nov 24. Weighed again about 2 pm. Captn. Newland of the St Lawrence, a friend of our skipper, gave the mate directions and we anchored in the Roads about 3 pm. I dressed and went on shore to Dr Martin's where I was very kindly recd. by Mrs D. Martin and several of their friends to some of whom I have letters of introduction. I learned with much regret during this visit that Mr Brooke would in all probability relinquish the attempt of forming a settlement at Sarawak as his very laudable exertions have not been supported by the home government. In a late affray whith the Natives several of his people were killed, amongst them Captn. Steward of the Ariel and the Chief Lieutenant of one of the Queen's Ships.

Nov 25. Engaged an appartment [sic] in the London Hotel at the rate of [illeg.] dollars per diem, have for a table companion Major Backhouse of Affgan [sic] celebrity. Dine in the evening at Dr Martin's and accompany Mr and Mrs Martin to their plantation in the evening. Nutmegs, Cloves Betel Nut, Cocoa Nut and many other tropical Fruit trees were in the greatest possible luxuriance.

Nov 26, 27, 28 and 29. Engaged delivering letters of introduction. On the evening of the 28th dined with Mr Blundell; on the 29th at Dr Martin's having previously called on the Honble the Resident councillor, H. E. the Governor and Dr Oxley, none of whom I found at home. On the evening however I recd. a very kind note from the Governor's Lady requesting me to dine there and inviting me to make Government House my home during my residence in Singapore. This I have declined as my stay will in all probability not exceed a few days.

Nov 30th. Mr Church the Resident Councillor called this morning and promised me every assistance in his power. After him Dr Oxley who did the same and invited me to dine with him on Monday. In the evening dined with the Sons of St Andrew, it being their patron's day.

Dec 1st Sunday. Mr Barnes of the Ariel in which I shall probably proceed to Borneo called, accompanied by a gentleman who is to command her, the Captn. being ill. In the evening dined at home and went to church for the first time since I left England; the service was conducted by the Rev Mr White.

Dec 2nd Monday. Called at Government House but found Col. Butterworth at his office. Planned an excursion with Dr Martin to Bukit Timah and in the evening dined with Mr Oxley who was kind enough to promise to accompany me on Thursday to a place where the Nepenthes are abundant.

Dec 3rd Tuesday. Called on Col. Butterworth at his office and found him exceedingly affable and anxious to assist me and invited me to dine with him in the evening which I accordingly did. We had a talk about plants and the Islands and countries of the East and from him I learned that it was not impossible to get to Siam, a place I should much like to see and which after having been to Borneo I may perhaps visit in preference to Java. But I fear the opportunities of sending home the plants would be very few. In the morning I had called on Mr White the Episcopal clergyman with Mr P_____ and whom I found enthusiastically fond of his garden. He had several Norfolk Island Pines and also the Moreton Bay species which he had brought from New Holland when on [a] visit there for the benefit of his health.

Dec 4 Wednesday. Learned from Mr Read of the firm of Messrs Johnston and Co. that there would be a Dutch vessel for Reiger in a day or two. I shall in consequence make no further engagements but endeavour to proceed in her as here it is considered madness to venture into the Jungle on acct. of the tigers which are said on an average to carry off one man per day, principally Chinese. The Ariel for Borneo sails tonight and I cannot go with her on acct. of the letters by the Septr. Mail not having yet arrived.

Dec 5th Thursday. Accompanied Mr Oxley to a small Island about three miles distant where the Nepenthes grew in abundance and in some parts of it out of the debris of the rock but in such situations they did not appear to thrive so well as when growing in a rich soil and in that moist position the two species (if species they be) which Mr C____ brought to England were both there but only that with the Bright shaped patches was in flower. Setting aside all peculiarities in the structure of the pitcher, the plant is well worth cultivating on acct. of the beauty of its flowers and the neatness of its habit. They did not grow tall and straggling as in cultivation but formed neat [? dry] bushes, the highest about 3 1/2 or 4 feet. I very much doubt the two being distinct as I gathered one plant on which the two kinds were upon one stem and another on which the one stem produced the large pitchers and the opposite the other kind. In another instance the two above mentioned and another with a hybrid appearance were in the same mass but I did not examine whether or not they proceeded from the same root. This last I shall send home; it is a glorious specimen. The Julia, Mr Brooke's Schooner, arrived today and I met Mr [William Bloomfield] Douglas,

Mr Brooke's Nephew, who commencing here in the town I went on board with and am quite delighted with his behaviour. In the evening dined with Mr Purvis having had five invitations for the evening.

Dec 6 Friday. Packed a case of pitcher plants and went to the town where I met Dr Oxley with whom I dined in the evening, having previously arranged with Mr Douglas a trip to Rhio in a cutter which a friend of his had promised to lend him.

Dec 7 Saturday. Started in the morning to 'Bukit Timah', a hill situated nearly in the centre of the Island and the highest one in the place. I had previously sent a pony-cart to bring back the plants I might collect but it rained so incessantly that I could get nothing. The jungle is moreover so thick that unless with people to cut through it with the finest weather it would be impossible to penetrate it. This is the rainy season here: well it may be called so for the water falls in floods every day. I fear we shall do very little at Rhio provided we do not have better weather. Dined in the evening with Dr Martin who had accompanied me during the day. The more I see of Dr and Mrs Martin the more closely I become attached to them. Indeed, had it not been for their kindness I don't know how I should have got on here at all. They have frequently invited me to stay at their House but having refused the Governor's invitation I could scarcely accept another without giving Col. Butterworth offence. I have however promised to do so on my return from Rhio.

Dec 8 Sunday. Went to Church in the morning and to Arthur's Seat in the evening to dine with Dr Little. It is about 7 miles in the country but as usual the rain prevented me from seeing whether the country was at all likely to produce any plants which would be valued in England. I however think not as it seemed to be tolerably cleared in that direction.

Dec 9 Monday. Left Singapore in the Cutter Foam for a visit to Rhio and the Islands on the route. We had rainy weather and squalls of wind at starting but more fair towards evening. About four pm anchored off Puli 'Mongsa' close to a reef of coral; went ashore in the boat but found nothing of interest. On the way [at a] small Island a few plants of a Cycas were growing in the jungle and the beach was protected by mangroves from the encroachments of the sea. It was low water and many varieties of coral were in great profusion.

Dec 10th Tuesday. Went on shore early in the morning to that part of Puli 'Battam' which is opposite to Puli Mongsa. Here we found a considerable Malay town on the Bank of a River, the Residence of a 'Rajah' who at this time was on a visit to another Rajah living near Rhio. Here was a very fine grove of Cocoa Nuts and many other fruits. We purchased a quantity and went across the river to the Chinese village which is situated on the opposite side, it is not so extensive neither is it so neat or well drained as that of the Malays. About Midday we got under weigh and anchored at night under the West point of 'Bintang', one of the Rhio Islands. It was blowing fresh and about 10 pm our cable gave way and we were obliged to get out to sea. We however again found soundings and brought up with a small kedge of 5 or 6 lbs weight in 9 fathoms water.

Dec 11 Wednesday. We were too far from the shore to land and as it had been raining without intermission all night and still was, we did not attempt it but went on to Rhio were [sic] we arrived in the evening all wet through and one of the crew ill with fever. I wonder much that there is no more sickness on board for we have scarcely been dry day or night since we left Singapore. At five o'clock pm a boat came off from the Dutch Resident to know who we were and what we wanted. I sent him my letters of introduction and the boat was not long in returning with an invitation for Mr Douglas and I to come ashore and spend the evening. This however we declined as we had no chance of reaching the Government House without getting wet through and this we could do on board without difficulty.

Dec 12 Thursday. Went ashore with Douglas at 8 am. The Resident took us during half an hour's cessation of rain to the fort which is on a hill commanding the Harbour. It is beautifully clean and in excellent condition. The Officers in [the] Garrisson [sic] pressed me to stay a few days, promising to show me everything of interest, but I told them I should return in the dry season. This I was encouraged to do the more as the Resident promised me every assistance, offering me a room in his house and people to assist me in penetrating the jungle. He further said that if I would give him an intimation previous to my next visit he would send one of the Dutch Gun-Boats to Singapore to bring me down as safe opportunities of passing from one place to the other are very rare while Malay and Chinese [? praus] are passing and repassing every day. At one pm we took leave of the Resident and his Lady and returned on board. We were soon under weigh and at dark came to an anchor opposite Puli 'Loban' near a village and a fisherman came off and we purchased from him a quantity of fish, that is to say 3 or 4 hundreds, for a Dollar.

Dec 13th Friday. Pulled ashore to two houses a little apart from the village. They were surrounded by Durians, Mangosteens and amongst Cocoa and Betel Nut trees. The houses appeared deserted as they were in a ruinous condition but two starving Dogs and some half dozen fowls soon convinced us that the inhabitants were not far off. We had brought our cooking apparatus ashore, so we after bathing breakfasted in a recently erected shed. Two Malays then came from the village and told us that the persons to whom the plantation belonged had gone out to fish. They however gave us Cocoa nuts and anything else we required [and] remained with us until we left about Midday. At night it blew strong and we could get no place to anchor in so that we were forced to keep under sail.

Dec 14th Saturday. In the morning at day light had soundings of nine fathoms on the Johore shoal; tacked immediately and stood off. About 12 pm arrived in Singapore Roads and anchored along-side the Julia. She has discharged her Antimony and will sail in four or five days for Borneo. When I came on shore I found all my things had been removed to Dr Martin's where I am to reside till I leave for Borneo.

Dec 15 Sunday. The heavy rain in the morning prevented me from getting to Church; in the evening the weather was more settled. Dined at Lieut. Elliot's of the Madras Engineers who has charge of the Observatory; Douglas is staying at his House.

Dec 16 Monday. Occupied during the morning in preparing 4 large cases of pitcher plants to send home by the Chieftain which sails tomorrow. I also gave Captn. Smith a bill of exchange written out twenty-one days after sight for 60£ (288 Dollars) on Home as I hear the Mail has gone to China and I must leave this [week] by the Julia.

Dec 17. Before breakfast rode to Dr Martin's plantations to see some pitcher plants. They were beautifully healthy and I shall send a case of them together with some Nutmegs and Mangosteens immediately home.

Dec 18 Wednesday. Butterfly catching in the morning. Called on the Governor. Dr Oxley and Mr White talked a good deal about Siam and think I shall go there by and by. In the afternoon went with Mr Little to

Mr Armstrong's plantation to look at some trees which we were informed were covered with Orchids. Lost ourselves in a Tigerish looking jungle and got out with great difficulty. Found however a new pitcher plant altogether different from any I had before seen. The cups are small and green but most delicately and beautifully formed; this will be an acquisition. Monkeys of several species and Kingfishers [were] very numerous. On returning to the Road how was I surprised to find Mr Myrtle in a palanquin accompanied by my brother [Stuart] whose ship had just arrived. He looks healthy and has grown considerably since I saw him [? 30] months ago. We dined at Dr Little's Bungalow and Stuart returned with me in the evening to town where he was obliged to go on board the Ship. N.B. I have today discovered the magnificent Acrostichum grande, a large mass was growing on a tree in Dr Oxley's garden and he was kind enough to show me a tree perfectly covered with it from which I must get the Governor's permission to extract it as it is in his property.

Dec 19th Thursday. Went to the place where I yesterday found the Nepenthes and was fortunate enough to find in a continuation of the same jungle another species with a beautifully neat habit. Stuart was with me and we were nearly exhausted having walked so much in the burning sun when we fell in with some Cocoa Nuts, the water from which much relieved us. My servant had sadly misbehaved during the day, otherwise it was our intention to have brought home a load of pitcher plants. In the evening dined with Mr Read.

Dec 20th Friday. Visited during the morning. Dined with Mr White the Clergyman at 4 pm and with Dr Martin and Stuart at 6 1/2 pm.

Dec 21. Cannot venture out on acct. of the very heavy rain. Dined with Stuart and accompanied him to the boat which took off to the Ship and I took leave of him. They sail to-morrow morning. How fortunate I consider myself in having thus had an opportunity of seeing a brother whom I may be a long time meeting again and considering the places I intend to visit I think the chances are almost against our ever seeing each other more. Others have however gone through greater perils and I may possibly too be preserved.

Dec 22nd Sunday. The rain of yesterday returned this morning. At 9 am I sent my boy to the Greyhound with a note to Stuart and at 11 am received one in return [that] they had nearly everything ready for sea

41

and were waiting only for the Captn's return from the shore. They sailed about 1 pm. The Mail via Calcutta came in but brought me no letters.

Dec 23rd Monday. Went plant hunting accompanied by Dr Martin and had a fine day. Brought home a quantity of two kinds of Nepenthes which were very luxuriant. One species climbing all over the trees, the other 4 to 5 feet high, both very beautiful. Coming home the Dr killing some snipes and Quails which were numerous. While in the jungle we were annoyed by Ants and Mosquitos. We killed a beautiful but very poisonous snake.

Dec 24th Thursday. Packed the plants collected yesterday and went out Butterfly catching.

Dec 25 Wednesday. 'Christmas day' has stolen upon me unawares. On Sunday last I had no idea that it was so near at hand. At home we look for it a month before it comes and the cold weather gives notice of its approach. Here, however, the weather is delightful such as our warmest summer days; the short mornings and evenings are beautifully cool. Spent the morning snipe shooting up to my knees in water the whole time and the day at Mr Little's Bungalow. In the evening dined with the Governor and Mrs Butterworth. This and yesterday have been beautifully fine days, perfectly dry and I hope we shall have a continuance of such weather.

Dec 26 Thursday. Called with Mrs Martin on Mrs Church, Mrs Purvis etc. at Sir John de Almeida's. Looked into the garden of the latter and saw some pretty things. Wished very much to see him but he was invisible through illness. I believe the Musa textilis from which Manilla cordage is made is grown in his garden. In the afternoon went to Arthur's Seat to look at some Orchidae and catch Butterflies but the heavy rain spoiled all.

Dec 27th Friday. Visiting again! Government Hill, Dr Oxley's and Rev Mr White's; planned an excursion with the latter for Monday next but do not expect much from it as Mrs White and their children are going. It is however to a new place and probably I may see things which may be worth while. Returning I saw Mrs Col. Butterworth's Birds, a pretty collection. Many Pigeons and Doves, including the large blue one with a crest from Java, a pair of Argus pheasants perfectly tame and two of a smaller species more beautiful. Many other Birds as Quails, Lories,

Paroquets and Cockatoos but in a very So-So kind of an aviary. I suppose wire work cannot be procured here or the Laths and Netting now employed would surely have given place to it. In the afternoon went to a new locality to look for plants but found nothing. Caught a few Butterflies and told a fellow I would give him two dollars for 40 plants of Acrostichum grande if he would penetrate the jungle for them as those of Government Hill are inaccessible and I have not observed them anywhere else. The man said he would bring them in three days. I sent my boy this morning with a Malay boat for Nepenthes. He has returned but gone home without showing himself or reporting progress. Am told that the Julia will sail on Saturday (tomorrow) week. The Ardasur sails tonight for Bombay to meet the February Mail for England. I have embraced the opportunity to send letters home.

Dec 28th Saturday. In the morning packed the Nepenthes collected by my Boy yesterday; they fill one case. After which, accompanied by Dr Martin, called on a Mr Scott who has the largest collection of fruit trees in the place and from whom I expect to get some for Shipment home by the Georgetown. Mr Scott unfortunately was from home. Caught a good many very pretty Butterflies and returned. I have been very melancholy these last few days, from what reason I cannot tell but think the disappointment experienced by having received no letters is the principal cause. Tonight I feel particularly afflicted and should not wonder if from this and exposure to the Sun combined I am 'laid on the shelf' for a day or two. Have been thinking lately a good deal of AD and I wish very much to hear of her. While writing I have my feet in hot-water and I must go to bed although it be not yet 8 o'clock.

Dec 29th Sunday. Went to Church though not very well. Came home and felt better during the afternoon. In the evening rode out to Captn. Scott's with Dr Martin but did not see the 'Harbour Master', he being from home. Rode around the plantation which would be a fine one were the Trees (Nutmegs and Cocoa Nuts) less crowded.

Dec 30th Monday. Attended a sale after the Malayan fashion to purchase some few things necessary for Borneo and packed a case of beautiful young Mangosteens for shipment to England. Four Malays have been brought in here charged with the murder of an Englishman and plunder of a coasting Schooner. It appears that the Schooner belonged to a person here named Duncan who had fitted her out and given in charge to a Captn. with instructions to make a trading voyage in the Straits amongst the Islands. After a considerable amount in dollars had been thus amassed the Malays rose on the Captn. and murdered him. They then put the valuables into the boat and scuttled the ship. They were seized by the Sultan of Lingin and forwarded to Singapore where they are now waiting judgement. They will assuredly be hung and well they deserve it. I have been disappointed today in the excursion planned by Mr White but start early tomorrow morning for the Interior in a different direction to any in which I have hitherto been.

Dec 31 Tuesday. In the morning early went to Serangoon [and] found the good 'padre' in bed and fast asleep. Went out and caught butterflies for an hour, returned and found no one up yet then went to look at Dr Almeida's Coffee plantation. Very poor indeed, it will not flourish here. After breakfast went I know not were [sic] but the jungle was full of young palms of 10 or 12 species. I bargained with some Malays to get me a hundred for 2 dollars. Found Orchidae on the Mangrove trees, as also the Acrostichum grande. In getting this last I was somewhat amused. Two Malays were cutting a plant from a tree when some black ants which had their nest there ran out and stung them fearfully. The two fellows leaped down and I could not prevail on them to venture up again. During the greater part of the time I was wading up to the middle in water but fortunately it was salt, otherwise I think such a Bath would have been by no means beneficial. Dined with Mr White and returned to town in the evening. During the day I had seen the Aeschynanthus before mentioned in flower. It is very beautiful, much superior to any of those already in England. At night went to a grand Ball and supper at Government Hill, this being the last night of 1844.

1845

Wednesday Jany 1st 1845. After leaving Gov. Hill this morning went first footing with three friends, A. Spottiswoode, J. Conolly and Harvey, a bottle of Whisky, one of Brandy and one of Cherry cordial. Served out every one from [illeg.] and made each drink wishing them at the same time a happy new year. After having made all our calls, wound up by filling the peons with spirits on the road home. Went to bed between four and five and was up again at 8. This being New Year's day was kept as a holy day. Dr Little's pony won the principal race. Sundry Gentlemen were 'drunk' but everything was properly conducted. In the evening dined at Dr Martin's with some of his friends [and] had other invitations.

Thursday Jany 2nd 1845. Employed during the morning in setting together some few things to take with me to Borneo as nothing in the way of eating and drinking can be purchased there. Called on Captn. Scott of [? H] and Co's Steam Ship 'Phlegethon' who has been much in Borneo. He promised to give me two letters of Introduction to Native Chiefs in Borneo proper [Brunei]. In the evening fell from the Horse which gained the Cup yesterday but fortunately was not seriously injured. I [was] turning him sharply at full speed when the strap of the stirrup on which was all my weight gave way as I fell on my head. Had there not been grass beneath me I should probably have been killed. As it is a black eye and slightly sprained wrist appear to be all the damages.

Jany 3rd Friday. Head this morning and throughout the day very 'shakey'. Employed in preparing for my voyage and in packing plants for home. Dined in the evening with Mr Myrtle of the firm of Messrs Armstrong and Co.

Jany 4th Saturday. Went to the Jungle to look for plants and succeeded in finding another species of Aeschynanthus perhaps more beautiful than the former. The leaves are very small and putescent but the flowers are very large with a scarlet corolla and large violet bell-shaped calyx. It will be a great acquisition. I also found one of the most magnificent plants I ever beheld, an Ixora with orange coloured flowers. The cyme is as large as a Hydrangea hortensis, the leaves nearly 11 inches long of beautiful graceful texture. Its habit appears to be that of a Combretum purpureum — but perhaps not quite so much of a climber. Should it be new at home and I can possibly get hold of it, will it not be valued? I could find neither seeds nor young plants though I searched most carefully. The Jungle in which I wandered today was composed mostly of palms, the hooked spines of which rendered the penetration of it very difficult. I saw many beautiful ferns and lamented that they should be so little valued at home. Orchidae were not to be found, indeed I wonder where they grow as I have never yet fallen in with them in abundance. In the evening I went Snipe Shooting and had good sport but in a most awful bog. For an hour I was wading in mud and water nearly breast high and could not find my way out. When I did I was perfectly exhausted, not having strength enough left to thrash three or 4 Klings [Indians] who would not give me some water. Dined at Dr Little's Bungalow with some friends of his.

Sunday Jany 5th. In the morning marked and otherwise prepared six cases to be sent to England by the Georgetown and Cucles, the former to sail in a day or two for Liverpool, the other for London in three weeks. In the evening dined at Mr F. Martin's, the Doctor's brother. All is now ready and the Schooner sails tomorrow morning and as I am informed by a note from the Captn at 8 pm.

Monday Jany 6th. Was accompanied on board the Julia by Dr Martin who left us when the Captn. came on board. Sailed at about half past nine with a steady breeze from NNE and worked beautifully out of the Straits till we came to the NE point of Bintang. The whole night was employed in beating around this.

Tuesday Jany 7. Steady wind from NNE. If it were something more to the Northward we should be all the better pleased but we need not look for that as this is the period when the NE Monsoon is at its height. Mr Mins who was mate under Douglas has command of the Schooner as the latter being in feeble health is to take the Ariel home.

During Wednesday and Thursday the NNE winds still continued and on the evening of the latter day we saw the coast of Borneo, that part which lies between Tanjong Api and Sambas. We have throughout been troubled by a head sea [and] the consequent uncomfortable pitching of the schooner. My Boy has been very sick and I believe sulky. He told me only the day before we sailed that he would not go with me unless I paid him 10 dollars per month. This I was compelled to do as I had no time to get another. He is afraid the Dyaks will kill him. I shall send him back with the Schooner.

Friday [Jany 10]. Very heavy squalls from NE ship pitching and rolling and 'up to all sorts of fun'. Chieftain's worst pranks nothing to this young Lady's (only 4 years old) playing. Had to work around Tanjong Api but found towards night that it would not do so ran for a small Island (St Pierre's) which bears nearly West from the Cape to see if we could find shelter. Let go the anchor in 17 fathoms hard bottom (broken coral and shells) but soon found she was driving. After some time and having given her all the chain she brought up about 3/4 mile from the Island. Rolled dreadfully all night, the wind having veered more to the Eastward and blowing very heavy the Island did not shelter us.

Saturday Jany 11th. Wind and sea as yesterday. Towards evening many birds resembling Noddies collected over the Island on which they finally alighted.

Sunday Jany 12th. Still very squally from the NE and consequently we are still at anchor. After very much solicitation I prevailed upon the Captn. to get the launch out to go ashore as the vegetation seemed so very luxuriant and I fancied that through the glass [telescope] I could see Orchidae on the trees. We had very much difficulty in landing on acct. of the rocky nature of the place. Indeed, the whole Island seems to be little else and one wonders how the trees which are so vigorous find nourishment. I however was disappointed respecting the Orchidae, the green appearance of the stems being attributable to Hoyas and other climbing plants which were in great abundance amongst many trees which I did not know. I observed in luxuriance the Barringtonia speciosa, a pandanus and a wild Plantain in fruit but not ripe to gather with the Birds Nest and other ferns but the inaccessible nature of the place and the number of centipedes made me turn again to the sea side on the rocks of which I found a great variety of shells Crabs etc. Having collected about a peck of the former when I returned on board they were handed to my servant with instructions to put them into a bucket and pour in some hot water to kill them that I might clean them after dinner. The fellow however thought he knew more about [them] and assisted by the Lascars they opened the Oysters [illeg.] etc., ate the animals and destroyed the shells. I shall not fail to take this crime into acct. when I settle in Borneo with my Boy for his other misdemeanours. Thus has my first collection of shells been disposed of and some of them I thought very beautiful. In cutting through the Jungle I had the misfortune to break a sword belonging to Dr Little and which together with a tent he had kindly lent me.

Monday Jany 13. Weather more moderate but still squally. Got under weigh at 7 AM and at 4 PM weather Tanjong Api with wind NNE. At 10 PM were Safely round Tanjong Dattoo [Datu].

Tuesday Jany 14. Weather fine and our course being SE the Wind fair at 7 AM were abreast of Tanjong Poe and spied the Brig Ariel just beating out. At 4 PM anchored about half way up the Sarawak River near the mouth of a tributary stream called the Quap. Left the Schooner in the Launch and accompanied the Captn. to Sarawak. Dined with Mr Brooke and a few Europeans residing here who received me very kindly. Returned to the Schooner at night.

Wednesday Jany 15th. Weighed at daylight and were carried by the tide being at the same time towed by the Longboat nearly to the town but we could not reach it as the tide had turned. This is a noble river navigable for ships of any tender up to the town which is 24 miles from the entrance of the River. The banks are beautifully diversified during the first 14 or 15 miles but after this though still pretty they are covered with thick Jungle in which innumerable monkeys find shelter. I observed many Orchidaeous plants and some palms. The Nibung and Nipah are very common. Altogether I think appearances are very promising and I hope to find a rich harvest.

Thursday Jany 16th. The tide carried the Schooner as far as the town which seems large but composed with few exceptions of Malay Houses built as is their custom on poles to prevent the tide from coming in to the Huts. Mr Brooke's House is built of wood on an eminence at the end of the town towards the sea. This, with Mr Williamson's the Interpreter and one built by Mr Steward, are the only houses which at all differ from ordinary Malay dwellings. Mr Brooke has promised me a room in a house which was built for him by the Rajah on his first arrival but it is at present occupied.

Jany 17th. From this day until Monday the 20th I slept on board the Schooner and dined with Mr Brooke as my room was not evacuated. On Monday I took possession and found during the night that the rats had a mind to dispute it with me. They are in great numbers and carry away everything that may chance to be left out. I have made several excursions in the Jungle and have found ten interesting plants, viz. a willow-leaved Ixora and a plant with cream coloured flowers something like a Banksia but the flowers are in terminal Bunches and the habit is somewhat climbing. Orchidae are rare in the Jungle but plentiful on the River bank. I must send to Singapore for a canoe as nothing can be done without one, there being no road but the River.

Tuesday Jany 21st. Weather Showery, thermometer at noon 82°. Employed in unpacking my boxes and getting my room in order.

Wednesday Jany 22. Weather as yesterday but with heavier rain. Could not get out. Dined with Mr Brooke. The conversation after dinner turned upon a Malay who had been caught during the day with property on his person which had been stolen from a house recently during the night. He is a native of Sambas or Pontianak, a vagrant character, and is strongly suspected of having been concerned in other burglaries. The Malay Law ordains that such an offender should lose his left hand but Mr Brooke is of course against mutilating his person but without a despotic use of his authority can scarcely do it. The native prejudices will not allow a man to be beaten and a fine this offender could not pay. Imprisonment is also contrary to Law so that there seems but little chance for the poor wretch. Mr Brooke however tends to mention corporal punishment to the Dattoo Patingi in order to save the man's hand but it is expected that the Malay Laws are something like those of the Medes and Persians: they alter not.

Thursday Jany 23. Very heavy rain during the greater part of the day. Mr Cruikshank and I went to the Dyak village which is situated on a creek a little way down the River. It is very different from those of the Malays, the Houses being built all under the same roof so that a large verandah is the road, on one side of which open the doors of the houses. In this verandah the Children play and the women work. They were principally employed in making mats which they do very beautifully but few families were now residing in the village, the remainder being at their padi plantations. The whole village is raised above the marsh in which it is built to the height of 4 or 5 feet upon posts not of hickory which the Malays use for this purpose but of some harder wood. These people are very fond of ornaments of beads and brass wire. One woman had her arms from the wrist to the elbows covered with massive rings of silver. Her neck was also encircled with about a dozen hoops of the

49

same metal. On returning from the village down the creek I saw many orchidaeous plants amongst them, two which I hope will prove new species of Dendrobium, Polypodium quercifolium is particularly among them. The prisoner I mentioned last night was privately examined today. He confessed to one robbery which indeed he could not deny but persists in saying that he knows nothing of the others. The natives will not hear of corporal castigation and Mr Brooke justly thinks that setting their own law aside for his authority will tend to weaken his influence among them. If possible he is determined to save the man's hand. Though this seems to us barbarous it was doubtless at its institution exceedingly judicious as crime amongst the natives is scarcely known, a case of theft not having occurred for 18 months. This is an idle scoundrel who has been offered work from several people but will not accept of it. He was formerly in the employ of a German Missionary who travelled overland from the Dutch settlements and discharged this man on his departure hence in the Ariel which we spoke of Tanjong Poe. After dinner the Rajah was waited upon by many of the Natives, both Malays and Dyaks. This is a nightly custom and Mr Brooke by his engaging manners and pleasing discourse cannot fail to render himself more beloved than he would be supposing they were debarred from his presence. One of the Dyaks who called this evening was a very good looking youth and very richly and tastefully dressed. They seemed to be a race capable of great improvement having good natural abilities and the present condition of those in the Sarawak territory compared with that which existed on Mr Brooke's arrival abundantly testifies that they embrace every opportunity which offers (Thermometer at noon 83°).

Friday Jany 24th. Employed during the day in writing letters to Singapore and one to my brother Stuart as the Schooner sails to-morrow. The Malay was tried at one o'clock and sentenced to have his hand cut off; this Mr Brooke remitted to three dozen lashes with the Schooner's Cat. Two thirds of them were accordingly given, the other dozen he was pardoned by Mr Brooke. He is to be expelled the Country. At the trial which took place in the Court House which was formerly that of one of the Rajahs, the Dattoo Patingi was dressed in a Company's uniform coat which he had purchased at the sale of the late Mr Steward's effects. The Levee in the evening was very badly attended but the weather was exceedingly wet which possibly was the reason.

Saturday Jany 25. The Schooner sailed this morning and with it I sent my boy with whom I have not been satisfied. On the day previous to our departure from Singapore he told me he would not go with me unless I paid him 10 dollars per month instead of 7 which in Singapore is considered high. I have moreover lost many things which as he has my keys must be attributed to his negligence. As a make shift Mr Ruppell has lent me a Dyak boy who cannot speak a word of English. I suppose I shall have to purchase a Dyak out of slavery as there is no other means of getting one here. This, too, will be an act of humanity and the boy receives but little wages until he has repaid his original cost. I am however devilish short of money and until the Schooner comes back would rather have nothing to do with it. Today I must get ready for an excursion up the Country which is intended for the beginning of the week provided the heavy rains do not create a fresh in the river sufficient to prevent us. Today I have partially concluded arrangements which I hope will materially contribute to my domestic comfort. The last I made of the kind were unfortunate but every thing looks better this time. After dinner we were visited, that is to say Mr Brooke was, and we were present by a party (said to be a deputation) of the Sakarran Dyaks who are the first that have come here peaceably. This is the piratical tribe which was so severely chastised by the Expedition [by Captain Keppel] of Octr last. They spoke of the dread in which they live of another visit and describe the horrors usually attendant upon such destructive ways. Their means of subsistence were entirely destroyed by their Fruit trees being cut down and the villages Forts and Houses were all burnt. They are now wandering in the woods without settled habitation and heartily wish for peace which Mr Brooke who is a most humane man will most cordially give them. They will of course submit to his authority as they live on his territory and a necessary preliminary will be that they abandon all piratical habits and practices. After they had retired our conversation turned upon Religion as connected with the Science of Geology from which soon became purely Theological. I find Mr Brooke's sentiments and my own approach each other considerably. We both believe in the Unity, Indivisibility, Omniscience, Omnipresence and Omnipotence of the

God of His Mercy. Direct me during life so that provided annihilation be the lot of man I may at least during my existence be of some service to my fellow creatures, but if this unending wisdom has prepared for the the disembodied soul regions of repose and bliss where sorrow never enters, where death can never intrude, where the Spirit freed from all that alloys even its most innocent pleasures here below and sharing its most transporting but transitory enjoyments shall spend an Eternity of unmingled happiness, may His Grace so guide me that after 'this mortal be swallowed up in immortality' I may find myself an inhabitant of those blissful regions. How happy is the lot of those who have already changed this life for that better one and how happy must they also be who live in the full assurance of sooner or later claiming seats beside them and finding themselves (by whatever means) accepted, welcomed, and beloved.

Sunday Jany 26. Very heavy rains all day. In the evening on Mr Brooke's hill I found an animal which of all others I have ever seen the most delighted me. It has on its body 36 beautifully luminous spots, that is to say 3 rows each consisting of 12. It has the power of suppressing the light from them at pleasure and gradually to cause it to reappear, sometimes those at the head and sometimes the lamps at the tail end of the animal becoming first illuminated. It is scarcely an inch long and its back very much resembles a centipede but tapers gradually from the middle towards the head and tail. It appears to have but six feet so that I think it must be a larva probably of some beetle (a fire-fly). The last segment of its body, its tail, it uses in pushing its body along. It is very active and one which suppressed its light I lost. I found them under a plantain; they were two in number but there were perhaps more amongst the dead leaves at the foot of the tree. The effect of the whole of its luminous spots is beyond conception beautiful. I do not recollect having read or heard of anything like it and you may conceive the pleasure with which I first beheld it. Oh! could they see it in England how would the admirers of Nature be delighted! After I have examined it by daylight I will preserve it to send home and search narrowly for more in the place where I discovered it.

Jany 27 Monday. The luminous animal is certainly the larva of a beetle; by daylight the only trace of its luminous spots is a faint streak where at night it is illuminated. During last night and till 12 this morning the very heavy rains continued. About that time it became comparatively dry, showers occasionally passing. I have engaged a Dyak to assist me in collecting who I am informed is thoroughly acquainted with the Jungle and we went together down the river accompanied by my boy to collect Orchidae. We tried the first creek we came to but it being low water we could not get up so we paddled for the Dyak Creek where I had previously seen abundance of plants. We were not disappointed: I found two Vandas and a Saccolabium, one of the former in flower beautiful and delightfully scented, 4 Dendrobiums, 3 of them new to me, both white flowers exceedingly pretty and a most elegant Fernandesia with large flesh colored and white flowers, the whole plant not more than 3 inches high. We soon filled our boat and after exploring a little further returned as we saw heavy rain clouds coming up. In the evening Mr Ruppell brought me a large Lizard which a boat had brought from some other river. I gave the man half a rupee for it and accompanied Mr Ruppell to the Boat to see an Orang Utang which was for sale. I bought it for 2 dollars; it seems very tame and I shall endeavour to render it more so as it is still young.

Jany 28 Tuesday. Very heavy rain during the night and a strong fresh in the River which will prevent us getting up the River. My Dyak went out and brought in flowers of some pretty things, amongst them two beautiful and distinct Ixoras. They have no seeds upon them at present but in the proper season I shall not fail to send them home.

Jany 29 Wednesday. Still raining heavily so I amused myself with drawing one of the Ixoras of yesterday. My man again went to the Woods and brought flowers of other beautiful plants. The most interesting is an Aeschynanthus whose flowers are in large bunches and each nearly three inches long; the calyx is as brightly colored as the corolla.

Jany 30. With Mr Ruppell's boat accompanied by my Dyak Tekei went down the River for Orchidae. Very showery weather but not such continuance of Rain as yesterday; with some difficulty on acct. of the lowness of the tide and muddiness of the River bed. We got sufficient by night to fill our little canoe as I could not find Suniat my boy when we left. I had double work at the paddles and returned very much tired. Two or three different Orchidae repaid me my trouble and compensated for my fatigue. I however immediately fell upon a bottle of claret which as I had nothing with me down the River was soon made away with, then dressed and went to dinner.

Friday Jany 31. Making ready for our River excursion there has been but little rain last night but the morning though fine promises plenty more. Mr Brooke, Cruikshank and I form the party. We go in two boats and the Chief of the tribe we shall visit goes with us in another boat or two accompanied by some of his people. About 1/4 past 4 pm we started and about 8 pm arrived at the Chinese village of Seniawan. The River to this place is broad and the banks beautifully diversified; sometimes we are surrounded by the thickest Jungle at others large plantations of paddy formed an agreeable relief to the eye. After dark around particular trees the fire-flies were in perfect clouds, presenting at a distance the appearance of phosphorically luminous atmosphere. On a nearer survey they looked like moving stars revolving round their favorite trees.

Saturday Feby 1st. We left Seniawan with the flood tide about 10 am and immediately passed a very large and apparently flourishing plantation of sugar cane belonging to the Chinaman at whose house we slept last night. He had also very fine Serih plants (Piper Betel), the leaves of these the natives chew with their Lime and Betel Nut. The sugar cane is sold for chewing also and of it all are fond. None is manufactured into sugar. The Banks of the River from Seniawan towards the interior are very steep but still during the last heavy fresh we could see traces of its having overflowed them all. The Chinaman told us that his house was flooded to the depth of 4 feet and I suppose it could not have risen less than 20 feet from its ordinary level. We passed many large tributary streams and the River is now much contracted; the Navigation for boats is rendered somewhat difficult on acct. of the large trees which have fallen across the stream and two or three small rapids render caution necessary when the water is low. Above our heads we have observed today several Dyak bridges; these are the most primitive I have seen. Two trees are selected which overhang the River and whose main branches approach nearest to each other. From these branches poles of three inches diameter are thrown across fastened together by the ends with rattans; these are again supported by the overhanging branches of the trees from which rattans [illeg.] are suspended. A rude railing is then formed by upright pieces of wood to which cross pieces are fastened and the whole is lashed together by Rattans to the poles which make the flooring, if it may be so called, of the bridge. Sometimes the rails are on both sides [but] more frequently on only one and the bridge is generally from 40 to 50 or 60 feet above the water. A little boy in our boat wittily observed when we came upon the first one that 'were he to see any crossing such a bridge we should certainly consider him a monkey'. The upper surface of the pole you walked upon is generally notched so that your path will probably be two and a half or three inches wide. Towards evening we arrived at the landing place belonging to the village of the Sauh Dyaks; from here we had to walk about a mile on notched poles such as I have described as used in the formation of the bridge. At this time it was not in all cases necessary for us to walk on them as the ground was dry, excepting which was very frequent when we came to a bog. Then had you lost your balance you would soon find your self up to the middle in mud. The village is situated on an almost perpendicular mountain of granite about 600 feet high and the ascent is formed of poles and large branches of trees notched deeper than for common walking. Before I reached the top of the Hill I was regularly knocked up. I felt giddy and thought I should have fainted and laid down on an old tree. The first thing which brought me my senses was our fat Dattoo, the Native Chief at Sarawak, with a little whelp of a son laughing as if to threaten the dissolution of their frames. I jumped up and with a desperate effort reached the top on which the village is situated but I still had to climb notched poles to get to the Head-Man's House which I could not do so I again gave up and Mr Brooke found me on his arrival stretched at full length in the first verandah. At last however I did get up and found my companions in the house belonging to the Chief. It in no way differs from the others like them; it contains but one room but they appear to be able to make shift when they have strangers for all the Houses communicate with each by doors inside. Besides this they are built of the same [? manner]. At Dinner the whole house was crowded with men, women and children, nor did they go away until we blew out the light. We then being all tired slept soundly until next day.

Feby 2. We are today going to a valley named Nawang in search of Deer but have promised the chief to return to-morrow when the whole tribe is to be assembled, a pig is to be killed and a feast to be made in honor of Mr Brooke's visit. At 11 am we started for Nawang with about 15 Dyaks marching behind us carrying our provision etc. Cruikshank soon left us behind and Mr Brooke and myself determined to take it easy. After half an hour's march under a burning sun I found we must halt to drink; we did so and found the bottles had been left behind. Sent a man for them and waited his return. We then passed on with the heat most oppressive over plains which had once been cultivated but were now overrun with Lallang grass on acct. of the terror in which the Dyaks have been living on acct. of the Sakarrans who were however licked last year. Some few years since the Sakarran Dyaks came and burnt their villages, destroyed their plantations and carried captive the women and children. On Mr Brooke's arrival they had scarcely a woman amongst them but partly by negotiation and partly by menace he has had the happiness of restoring them to their families. After leaving the bogs and Lallang grounds our road lay occasionally through Jungle but more frequently through beautiful valleys and after having crossed a ridge of low hills the scene was magnificent. Before us lay a succession of valleys, some which were well cultivated, bound in on all sides by beautiful hills covered with immense forests to their very summits. Here and there the dark green monotonous foliage was beautifully relived by Cocoa nut, Sago and Gomuti palm. Dyak Houses were scattered at the feet of the hills and river wound through the whole extent of the valley. We arrived at Nawang which is a well cultivated valley about 4 pm and soon after loaded our guns and went in search of deer. We saw several but could [not] get near enough to shoot till nearly dark when suddenly, after having separated from Mr Brooke, my man pointed to a deer standing in the Lallang. I could not see his head and neck. I fired at the latter when out of the same grass jumped four other deer. I fired again but to my extreme mortification I found I had killed neither. Neither Mr Brooke or Cruikshank were more successful and on our return we found we had nothing but fowls and bread and cheese for dinner. I had gone shooting with bare feet and found them now devilishly sore, the grass having cut them, leeches sucked the blood from them and the ants seriously bitten them. Mr Brooke however was worse off; he had forgotten to bring any shoes and walked from Sauh to Nawang and back without any. The day of our return to Nawang was dreadfully hot and the paths were like plates of heated iron. All the stagnant water we came to was so hot as precisely to resemble a warm bath and was just as much as one could keep his feet in with comfort.

Feby 3. Arrived at the mountain about 3 pm; at five we were summoned to the feast. I was asleep and did not see the beginning of it but I believe it was some prayer or incantation invoking heaven to bless their labours and give them good crops of rice, Sago etc. The feast consisting of rice boiled and put into bamboos about 2 feet long. After it had been spread before us it was taken away and divided equally amongst all the houses in the village, a fowl was disposed of as far as it would go in like manner and then the pig. This pig had been bought by the paddy belonging to the tribe from its owner for this particular occasion. It was now distributed amongst the people. After this came a dance performed by five of the oldest and most honorable men in the tribe. They first saluted Mr Brooke and the Europeans then the Malay Chief. Their music was a gong, their dancing slow and not inelegant but too monotonous. On these occasions, that is to say peace conferences, the old men dance. On a war feast the young men and in some tribes the girls perform this part of the ceremony. This tribe never allows the women to share in these feasts otherwise than as spectators. After the dance there was a long speech complimentary to the 'Tuan Besar' during which and the dance the toddy extracted from the Gommiti palm was freely handed to the old men and some of them during the speeches began to get noisy. We then returned but they kept up the feast outside the verandah all night.

Feby 4th Tuesday. At 9 am we commenced the descent of the mountain. On our return we soon reached the boats and arrived safely at Sarawak or Cu-chin [Kuching] at 6 1/2 in the evening. It rained from the moment we set foot in the boats till 8 at night but we were perfectly sheltered by the Kajangs [thatch] from the weather.

Feby 5th to the 12 were spent in collecting and in some measure delineating plants and [? birds]. Very successful with the former but the latter are difficult to obtain. I have seen in flower some beautiful Orchidae on a Banda tree which I suppose are Coelogyne though I can but guess are really magnificent. Some Dendrobiums etc. are very pretty. The Vandeae are generally found in but small patches: this must account for my not being able to send them home in large quantities. The kinds with pseudo bulbs are much more abundant. From what I have observed of the habits of Orchidae in general they do not materially differ from other countries; they prefer moss covered (but not dead) trees with but little foliage. Such, provided they overhang a creek, are sure to be covered with them. On one occasion I found 10 or 12 species on one tree. In the timber forests they are rare but probably the few found there will prove of a different character from those inhabiting the banks of streams.

Feby 13 Thursday. Weather very hot. I merely entered into the Jungle close to the House with my gun [and] shot 3 or 4 specimens of Birds which I skinned. One, a small thrush, has the prettiest note I have heard in this country. Mr Ruppell bought for me from a native a beautiful small Boa constrictor which I have put in spirits. I myself purchased a young Hornbill for a Dollar which I intend to endeavour to bring up. It eats birds with avidity and seems very healthy.

Tuesday Feby 18. Went up the River to collect plants and shoot birds. Fell in with some Bandas etc. and shot 19 of the beautiful Honey birds. These gorgeous little creatures of [illeg.] [are] much less beautiful than the Humming Birds of the Western World. In size they are about the same, in colours equally they possess the beautiful metallic lustre. They frequent high moss covered trees and build their nests of the sides of such. One which I found on a tree that had been felled was fixed by the bottom to a very small branch on which a young Vanda was growing, its roots having passed through the bottom of the next fixed it more firmly to the tree. On our return we saw two gentlemen walking with Mr Brooke on the lawn before his House. As we could see nothing of any ship we could not conceive who it could be but were not long in suspense. One was Captn. Bethune R.N., the other Mr Wise from London. The Captn. is sent out by Government on a diplomatic mission and is the bearer of an appointment of Mr Brooke and connected with the Island of Labwan which [Government] are going to buy from the Sultan if it is found available as a Harbour for shipping. The Captn. and Mr Wise had been brought over by H. M. Steamship Driver which was lying about 4 mile down the river.

Wednesday Feby 19th. Went up the River with the 2nd Lieutenant, the Dr., Gunnery Officer and another from the River to the Dyak paddy Fields on the left hand side. Dined with the same party, Mr Wise Captn. Bethune etc. Was invited by the latter to accompany them to Borneo but cannot very well go at this time as there are but about two months more in which I can send home plants so that they may arrive in the Summer time. After June I can go anywhere.

Thursday Feby 20th. Was again invited to go and look at the vegetation of Labuan but finding that in the Steamer I could bring no live plants home again declined though very anxious to go.

Friday Feby 21st. The Borneo friends left us about noon to join the Steamer which left the river soon after. Recently I have attained many good Orchidae. Today after Banda with purely white flowers and very fragrant. Another marked with cinnamon from a yellow ground with a purple and white labellum, also fragrant, both growing on trees in the Creeks; neither abundant but the white one most frequent.

Saturday 22. Staid at home unwell.

Mar 5th. The Schooner arrived putting me in possession of letters from which I have so long and so anxiously expected. All my family are well but little James has been very ill though he had recovered when the letters left. I would to God I could say so of another who though no relative is perhaps more dear to me but that the next accounts I receive of her health may be more favorable I shall sincerely pray and hope. In other respects everything the 3 mails now to hand contain is good; may the news they in future bring never be less so.

Saturday 8. Wrote letters home, one to the [illeg.] of England respecting matters of the most vital importance in which I trust everything may be guided for the best by the powers above us. But I can write nothing upon this subject; with difficulty I wrote what was unavoidable. I cannot repeat it here.

Saturday 15. The Schooner dropped down the rocks above a mile below the town. She is deeply laden with Antimony and looks beautiful. She has on board 4 very large cases of valuable Orchidae packed close and with them I have sent the only glass case I had full of Vandas etc., 8 species of these and some others all first class.

Sunday 16. Steamer Driver arrived from Borneo at the mouth of the Santabong where she is to take on board Mr Brooke accompanied by Mr Wise. The 2nd Lieut., the Gunnery officer and the mate came in the cutter. With the exception of the latter all will remain here till the Steamer leaves. Wednesday the 19th the party left us to join the Driver which sails tomorrow morning. I forgot to mention that Captn. Bethune sent me specimens of the 'Upas' tree which is found in Borneo and where he had procured them many strange tales are told of this tree but I will refrain repeating them till I myself see the plant.

Saturday 29th. Borrowed a small boat to accompany our people to the river 'Tabor' to see 'Tuba' fishing. The River is I believe a branch or mouth of the Santabong which itself is one of the outlets of the Sarawak. At low water the 'Tuba' having been bruised in the boats with mallets and water thrown on occasionally during the process, the juice of the nut mixed with the water forms a soapy looking fluid which collects in the bottom of the boat. A prayer is said and the signal given when all the boats begin bailing out the Tuba into the stream. In about 10 or 15 minutes a few fish came floating on the surface and were immediately speared by men standing in the head of each boat. They then came gradually for two or three hours, the largest fish not until the latter period had elapsed. Some of these were enormous fish. The scene is very exciting and our whole population went to see it; there were not less than a hundred boats on the spot. At last however the flood tide spoilt the sport and we returned on the evening of Sunday having about 30 miles to paddle.

April 2. I have this month sadly neglected my log but as little worth notice occurred during it it is of the less consequence. I made one excursion to Suntah up the Sarawak River where Mr Brooke has a small house and plantation. About 14 miles above this town of 'Kuching' or 'Sarawak' the river divides, or rather two Rivers join. That on the left hand is called 'Sarawak', that on the right in ascending is the 'Seniawan'. About 10 miles up the Sarawak from 'Lidah Tanah', the name of the Land at the junction on the left hand, is a small stream called the 'Suntah'. Two miles up this River is the estate belonging to Mr Brooke before spoken of. The Sarawak River in fine weather is the most delightfully pleasing; one can imagine its waters are clear as crystal. Its banks are clothed with trees of the most graceful appearance which at the time of my visit were in many instances in bloom. One in particular was so gorgeously beautiful that it must be mentioned. Its foliage was of the darkest shade and thousands of bunches of flowers of a large size and of the most beautifully bright lilac color made it form one of the most brilliant and magnificent objects of this very beautiful river. In two or three places I was attracted to the banks by plants of a species of Clerodendron; a most superb plant, it grew in bushes about 5 or 6 feet high, each branch of which was termined by a bunch eighteen inches long of crimson flowers. In the cleared spots a beautiful tree fern was occasionally seen and only a little house surrounded by the fruit tree of the tropics reminded us of what we might otherwise have easily forgotten: that man had penetrated here. On a beautifully still and moon light evening the beauties of this place are even more enjoyed. As we lay on the bosom of the stream I fancied myself an intruder of this delightful solitude, nothing at the place we staid for the night save a solitary Betel nut tree. The remnants probably of a Garden long overgrown with jungle was there to remind us of the proximity of any of our race. All seemed to belong to the beasts of the primeval forests on each side of us. Through the night the occasional shriek of a small or the more disagreeable scream of a large monkey and the melancholy note of an owl in a neighbouring tree were the only interruptions to [the] death like stillness of the scene. One cannot describe one's feelings in such a delightful solitude but at the moment I thought that nothing would be more delightful than to live and die a hermit there.

The Suntah river is noted for the Diamonds found in its bed. They used to be procured by washing the soil brought down by the heavy rains. At present they are not worked as Mr Brooke's preparations which had cost a great deal of money and were intended to turn our part of the River into a mere channel were washed away. Mr Brooke has had so much to do since that he cannot yet again begin and no native is master of sufficient capital. The Diamond workers of Banjarmassim wish to come but Mr Brooke I believe does not intend to introduce them into his settlement. At the Back of the house at Suntah, or rather at the end of it on the opposite side of the River in the Jungle, is a remarkable mountain. The face of it opposite to Suntah is every where perpendicular as a wall, it is many hundred feet high and composed I believe of limestone. On the other side it is accessible though very steep at its foot on the steep side the Dyaks of the 'Sampro' tribe have a small village; their Paddy Farms are scattered around it. The Nutmegs are planted on this estate by way of experiment; no care was taken of them, the weeds were for a very long time allowed to surround them, but notwithstanding all these disadvantages they have flourishing and are now growing magnificently. Cocoa Nuts, Betel Nut, sugar cane, Pink sweet potatoes and everything else which has been [planted] grow beautifully. In the woods I caught some beautiful butterflies and shot some pretty birds. A magnificent hawk larger than a falcon escaped me on account of the smallness of the shot with which I fired at him. His feathers flew in all directions but he succeeded in crossing the River which was there too deep for me to ford. A beautiful species of Trichomanes and a pretty Hymenophyllum were growing, the one on rocks in the Jungle the other on the trees overhanging the river. A curious cockscomb was in the paddy fields and many plants in flower on the banks. With the exception of the cockscomb I was too early for any seeds. Mr Brooke before leaving this [country] for Singapore after his return from Borneo gave instructions to Mr Williamson the Interpreter to carry a letter to the Dattoo Patingi Abdul Rahman at Serikei. On Wednesday the 16th of April he left this in the 'Buaya' War Boat of 70 feet length and a complement of 50 men accompanied by our Dattoo Banda and myself. Sherip Hussein, one of our chief men, commanded a smaller boat with about 20 men which was to go with us. During the night we lay off the fishing village at the Maratabas entrance of our village. Made sail the next day at noon and in the evening entered a small river or rather creek where [there were] a few fishermen. Here we remained for the night. Early on the morning of Friday we got under weigh and passing the 'Batang Lupar' and 'Saribas' Rivers together with others of less note, towards the afternoon we entered the 'Kalaka' River and about 4 pm anchored off the town near the House of Tuan 'Mulano' who holds the Country under the Patingi of Serikei the whole coast and as far as we could see inland. After having passed the 'Batang Lupar' was exceedingly low and covered with the thickest Jungle to the water's edge. Where the sea had formed a beach 'Aroo' trees (Casuarina littoria) are always found but when mud forms the boundary of the waters Mangrove is the first vegetation and immediately behind them the trees of the primeval forests. The Country between our River and the Batang Lupar including 'Samarhang', 'Sadong' and 'Lingii', appeared to be beautifully studded with mountains not disposed in ranges but for the most part solitary. Soon after our arrival we waited upon the old Priest Mulano (is an Arab and a priest) in a tumble down kind of reception room. He was delighted to see Williamson (whom he knew) and the whole room was filled with the inhabitants of the place. Tea and native cigars formed of Bornen tobacco very tightly rolled up in the leaf of the Nypah and tied with silk were placed before us and a long conversation was carried on of which to my misfortune I understood not a word. After about 2 hours 'Bechara' [talk] we returned to the boat, the old gentlemen accompanying us to the ladder of his wharf which was a mark of his most distinguished consideration. As we were descending to our boat the tide having quite gone out our noses were very disagreeably affected by the effluvia from the mud bank on which the town is built. As is the case with many Malay towns, the Houses are built on nibongs [stilts] in the water so that at high water in Spring tides their floors are but little raised above it and at low water the bank is dry. One would think that such a situation would be very unhealthy but it does not appear to be the fact as the inhabitants of those towns are similar in all respects to the others. One thing which always strikes one forcibly on seeing a Malay town is the apparently enormous multitude of children in comparison with the number of houses. You see them everywhere, in the water, in the boats, and if there be any of the dry land. When you enquire you find generally that each family in reality has but few children but as three or four families live in the same house the number of houses is necessary less than it otherwise would be and hence the disparity of numbers between the children and the houses. I have said thus much because one hears from strangers every day that the Malay women bear great numbers of children, whereas the contrary is most certainly the fact.

In the evening we were waited upon by Pangiran Illudin and Nakodha Seraddin. The former is sent by the Rajah Mood al [Muda] Hassim to collect tribute. The latter left Sarawak long ago with his praus laden heavily with the Rajah's property. They are two great rogues and are here rallying the people of the place who are afraid of them and the Dyaks in the Interior. The contents of Mr Brooke's letter to the Patingi is to inform him that all demands made by Borneons in the name of the Rajah are to be resisted if unreasonable and Mulano knowing this sent off to night to beg of Williamson to open the letter and read it to the assembled inhabitants of the place. This is to be done to-morrow.

Sunday in the Morning we again visited Mulano. Our Dattoo was prettily dressed in a jacket of blue cloth of fine texture plentifully embroided [sic] with gold. His Trousers were of crimson silk with gold thread worked into it and very prettily arranged. His Sarong was of the finest Sarawak make and his Kris mounted with gold. The handkerchief on his head was silk and gold threaded with a broad border of gold and nicely put on. Altogether, he being a young man of good features and slight figure, he looked remarkably well. He is the son of the 'Dattoo Patingi Alli' who was considered the bravest man in the country and was unfortunately killed at Sakarran in the same boat with poor Steward. Pangiran Illudin was there but [illeg.] though twice summoned to hear the letter would not come. I proposed to Williamson to take half a dozen of our boat's crew and fetch the rascal but it did not meet his approbation, so having waited for him an hour before the whole of the people and being then defeated of our purpose the letter was read. Every one present was pleased with it with perhaps the exception of Illudin. He however behaved as if he entirely approved of the contents though at that very moment his people were acting in opposition to Mr Brooke's expressed wish and he before our arrival had sent a 'Vertanah' (a musical instrument) up the Country to the Dyaks for which he demanded 40 $ in the name of the Rajah and when he found the Dyaks could not pay him money he requested or rather ordered them to pay in Paddy or Rice to the value I believe of upwards of a hundred reals. Of this we were told by all the Residents we saw and when Paddy [sic] was accused of it by Williamson before the assembled people he denied it without a wink and challenged any one to say to his face that he had done so. Of all the people not one had courage to accuse so Williamson could do nothing but warn him to leave the place as soon as possible. To do this however he will assuredly be in no hurry. There are those people terrified at the very name of Pangiran (it is the title of the hereditary nobles of Borneo) and dare not resist his demands, however preposterous or iniquitous they be. After having finished the conversation we left the priests [and] returned to our boats. I took my gun and after many attempts on the part of the natives to dissuade me, as they said on acct. of the Seribas Dyaks but as I believe on acct. of them thinking that I had some other objective than my avowed one (shooting birds), proceeded up the River in a small boat to see the Country. We went about 6 miles, five of which was through a Nypah Marsh of immense extent regularly overflown at high water. We passed several salt manufacturing establishments for which Kalaka is celebrated. It is made from the burnt ashes of Nypah leaves and though much used by Malays for salting their fish is but sorry stuff and abominably dirty in color. the latter mile of our puli was through Jungle with the banks bordered by Nypah. I could find but one place where it was possible to land and then in soft mud. In the Jungle from this landing place the Inhabitants of the Town get their supply of fresh water; it is very dirty and in no respect good, at least when it reaches the town six miles below. On my arrival at the Boat I found a dinner which had been sent on board by Mulano and soon after another arrived from Illudin. They consisted principally of sweetmeats and cakes made of Rice flour, a kind of custard and fowl dried rather than cooked over a slow fire. Spoons prettily made out of the all-serviceable Nipah leaf accompanied the Dishes. In the evening many people waited on Williamson in the boat and some sent presents. Mulano sent several fowls and ducks together with a sarong for each of us of the finest description manufactured by the Malays. In the Manufacture of this kind of cloth so much worn the inhabitants of Kalaka and Serikei excell all others along the coast, the cloth being much finer of course fetches a higher price. Pangiran Illudin had the impudence to send Williamson a Sarong in the evening which he accepted. We left about 3 o'clock on the following morning when Illudin saluted us with a lot of guns, thus wasting powder which did not belong to him and would nave been invaluable to the owner. During the morning of Monday the 21st we stood along the coast eastward with a fine breeze till we entered the mouth or rather the principal mouth of the Rajang. There are other mouths but they pass by different names. On doubling the point of land on the Western side of the entrance we had a view of the fishing village of Rajang. On most of the houses was displayed a white flag as a token either of fear or welcome. It could scarcely be the former as they had for some days been apprized of our coming. We passed the village which appeared to consist of very large and high houses with a fresh breeze and sailed about 20 miles up. The Stream is very broad with a strong tide running; the banks are well wooded and pretty. Here the wind failing, the ebb tide forced us to anchor about 6 1/2 pm. About 10 miles further up the Serikei River branches off to the right; this distance we accomplished early in the morning of Tuesday so that by 7 am we were laying just below the town. On our arrival there we sent to inform the Patingi Abdul Rahman who requested us to wait a little while that we might be properly conducted into the town. About 9 am while we were dressing and waiting for our breakfast a crowd of large boats were seen coming to the place where we lay. We were told it was the Dattoo himself and in a few minutes ascertained that such was the fact. He was in a boat pulled by about 20 men and followed by about 20 other boat which when he came alongside of us arranged themselves under the left hand bank of the River. He cannot walk on account of a very bad leg and his house having been burnt down in a fire which destroyed a great part of the town some time since. He now never leaves his boat. As soon as we signified to the chief that we were ready to proceed up the Town a signal was made from his boat to those in shore on which they immediately hooked themselves on to our two boats and towed us through the Town till we came opposite the reception room. During the whole time gongs were beating, flags flying, guns firing and nothing which could possibly give us an idea of the estimation in which the Patingi held Mr Brooke was omitted. All this was very gratifying to a well wishing of our 'Tuan Besar', considering that it was done by a Chief who himself had never seen an European before and none but those of his people who are [illeg.] and sailors in the praus could boast of being in advance of their chief. Every house was crowded with women and children all anxious to have a sight of our pale faces. The boats left us opposite the 'Rumah Bechara' or Court House of the Town which is situated on a low hill and surrounded by fruit trees. This, with the exception of one about the same size on which the Patingi is building a House, is the only ground which is raised much above high water mark in the neighbourhood of the Town, the remainder being flat and apparently swampy. After waiting for some time that the Court House might be fitted up for the reception of Mr Brooke's letter, we were informed by some of the principal people that every thing was ready for the receiption of the letter. It was placed in a brass kind of waiter with a [illeg.] two of the prettiest cloths they had embroidered with gold and tinsel. One of the persons appointed to conduct it then placed the vase on his head and carrying in this manner was followed by Williamson and myself together with three or four hundred of the natives to the room were it was to be recd. This walk to the House was decorated on each side with flags of gaudy colors and guns were fired all the time we remained on shore. The letter and its attendants soon reached their destination and the bearer of the precious scrap to which all this respect was paid sat down at the distance of four yards before a chair at the upper end of the room and placed the letter before him. Sachassan mats of beautiful workmanship were placed on the right of the chair for us and two or three of the most distinguished inhabitants the others all sat at the back of the letter Bearer. The room was hung with cloth of various colors and nothing seemed to be wanting but the presence of the chief. I was anxiously watching the hangings expecting him every moment to enter, as at that time I had no idea of his illness but I was presently undeceived by his nephew, a fine looking young man dressed in satins and silks adorned with abundance of Gold and who sat on the same mat with us, asking Williamson if the letter might be opened. On his giving assent the person who had carried the letter on his knees with his head bent to the ground placed the salver containing it upon the chair with many tokens of respect intended as I suppose both for the letter and the chair. A person then shuffled to the chair on his hands and took the letter which was at last opened and read to his honor the chair and the assembled multitude. A hum of approbation succeeded but few remarks were made as though it had been read to the Patingi's representative; the chief himself had not heard the contents. Now the chair about which all this ceremony was used was, as 'Boz' would say neither a new chair nor a good chair, but a very old chair and very sad chair, so small that had the chief himself occupied it it had certainly broken down. It was made of common wood and very dirty but covered with very pretty and very clean mats. An old box was placed at the foot of it as a footstool; this also was covered so that their imperfections [illeg.] it from the generality of observers but a fresh breeze which raised one of the mats gave me an insight to the mysteries beneath. After Williamson had made some comments on Mr Brooke's intention in sending this 'friendly epistle' (the purport of which was to explain his wishes respecting the treatment of unreasonable demands from Borneo people) we adjourned to the Boat in which the 'Patingi' lived. After a good deal of trouble occasioned by walking on bits of branches of trees and leaving the mud we at last reached his boat. I had fallen from one of the sticks and landed in a puddle of dirty water up to my knees; this as I was dressed entirely in white had a good effect. The Dattoo recd. us very kindly; he had been anxiously waiting for our visit and having read the letter told us how much he was delighted with Mr Brooke for having sent [it] and more especially so as Williamson accompanied it. He is a stout man of 50 probably about the height of the middle stature of Europeans, of an intelligent countenance but spoilt by two or three horridly long ugly hairs which grow from his neck much below the chin and which he appears to cultivate with great care. During our interview he ate an enormous quantity of Sirih so that I have never before seen anyone who at that could at all equal him. He was incessantly supplying his mouth afresh. He soon got into a long conversation with Williamson respecting the benefit it would be to his country could he introduce all the regulations Mr Brooke had done into Sarawak and expressed his anxiety to do so but at present many things interfere to prevent him, particularly the fear in which he lives of the Seribas Dyaks whose territories and those of the Kyans, another warlike race, join his in the interior. We were told that both in his River and Kalika small boats are constantly cut off up the River. It is done in this way: the Dyaks are in shore, their boat being concealed, and when the Malays pass in a small boat the Dyaks push off and without asking questions take the heads from those in it which they carry up the country in triumph, no plunder being so much esteemed by them as the heads of their enemies. The conversation lasted about two hours after which we returned to our own boats. As we were leaving, Williamson having shaken hands with the chief, I turned round to take up my hat and cane. The Dattoo saw this and fancied I was going to leave without any ceremony on which he hauled me to him by my left and heartily shook my right — his example was followed — 10 or 12 others who were of consideration sufficient to be admitted into the boat. In the evening I got a small boat to take me up the River to see the rest of the Town. The Houses on the right hand side are built on posts of hard wood about 16 feet from the ground. I landed at one and found it built precisely after the manner of Dyak houses, that is to say it contains many houses under one roof. In front of them is a verandah used by the inhabitants for cleaning their rice, making mats, baskets etc. These Houses belong to that tribe of the Melanoan race (which is a subdivision of the inhabitants of the Island as Dyaks, Kyans and Malays are others) called Rejang from their village being at the mouth of the Rejang River but this village is now mainly used as their residence as it would appear at certain seasons when very nearly all the inhabitants (Rejangs) leave Serikei to fish. This I was told by their chief who is named 'Gelang' and whose house it was I accidentally entered was the case during my visit and it was the reason that the greater number of these houses or rather villages were altogether without tenants. I also learned from Galong that the reason of the height of this house was their fear of the 'Seribas' Dyaks who are in the habit of calling in the night and running spears through the openings of the 'Lenteis', of which the flooring is formed, into the bodies of the sleepers above. There are about a dozen and a half of these villages and the houses they contain are of very much better construction than any Dyak habitations I have yet seen and few Malay houses are superior to those I saw. In their dress they (the people not the houses) in no way differ from Malays and the chief told me that he was a Mahomedan; probably the whole of this tribe have been converted to Islamism as they have been so long with the Malays but the Melanoan race in general are I am told similar to the Dyaks and possess but vague notions of Religion or the existence of a God. The opposite bank of the River is occupied entirely by Malays; their houses are very numerous and thickly inhabited and I should say that the whole population including the Saboo Dyaks who occupy two villages built like those of the Rejangs at the entrance of the town cannot be less than from 12 to 14,000. The Suboo Dyaks [illeg.] that particular kind of parang which goes by the name of 'Ilang'; its blade is convex on one side and concave on the other, about 20 inches long. I should fancy it rather an ineffective weapon as from its particular construction it can only be used from right to left. It is used as far as we can learn by the whole of the Kayan tribes (who inhabit the centre of the Island) and from what one had heard we suppose it to be peculiar to these people. The haft or handle is generally nicely formed and ornamented with the tufts of human hair generally dyed and the sheath is of wood ornamented with the strips of the skin of the civet cat and in many instances with carved work. On one side is attached a smaller sheath in which a knife is usually carried; it is frequently fastened to the body by a belt, one end of which has a loop and to the other is attached part of the beak of a hornbill or some other large bird which, being passed through the loop and slightly twisted, serves as a buckle. From this generally depend two or three tassels or other ornaments. I noticed that the Dyaks rarely laid them aside and swords and spears were seen in every boat however small or short the journey they had to perform. This would seem to argue an idea of insecurity which from the size of the place can scarcely exist to any great extent. The Country round the town for the distance of two miles on every side is clear of Jungle; this is probably done [so] that an enemy attacking them by land may be seen ere he can approach the town sufficiently near to fire it but it has at least in some place been cultivated with padi. I saw no fruit trees near excepting on the hill before mentioned. From the number of praus we saw of a very large size hauled up on the shore (about 30) it would appear that these people carry on a great trade and on enquiry of Williamson I fond such to be the case. They export Rice and Sago, Cloth, Bees Wax and paddy to Sarawak, the Islands and other places along the coast for which they get many things they want as Pottery, Iron, Handkerchiefs, Red Cloths etc. It is probably that some of their boats go to Singapore. Gold is found in the country but I do not know in what quantity. On returning to the Boat I had to pass the new House the Patingi is building. It will be one of the largest and most substantial Malay Houses I ever saw; all the posts are of 'Bilian', the wood called by Europeans from its hardness and durability Ironwood. It is built on a slight elevation and commands a view of the greater and most populous part of the Town. In the evening several people came off to the boat, some to see if they could do any trade with our people, others to talk to Williamson and not a few of these last wished to leave the Country and go to Sarawak but as they cannot do this without the Patingi's leave and he is naturally averse to it Williamson very rightly gave them no encouragement. The number of persons who are flocking to Sarawak at every opportunity is astonishing; whole tribes of Dyaks and families of Malays. At this time there is a tribe of Dyaks belonging to the Dutch settlement of Sambas anxious to come here and have sent their Orang Kaya to see if Mr Brooke will give them a settlement.

Wednesday. About 10 am we went to the 'Patingi'; during this visit the conversation was turned by the Patingi to European Nations and their customs and manufactures. The only two people he had heard of were the Dutch and English; he asked us abundance of questions relating to each, particularly respecting steam-ships of which had heard, they having passed the mouth of the Rejang several times on their way to Borneo. He was perfectly astonished and abundantly expressed his amazement, which the Malays seldom do, when told that carriages on land could travel by steam much faster than a ship at seas. He asked how we made so much cloth and many other things and when constantly answered 'by the power of steam' it seemed utterly to pass his comprehension. He then spoke of our cities, our country and its population, of our wars and relations with other nations and when our late wars were mentioned to him [and] the number of men killed in the great battles he figured that it was more men than their country could produce. This must certainly have surprised him as the wars he hears of are in general very bloodless, unless when they succeed in surprising a town or Dyak village then 3 or 400 heads is the very outside that I ever heard had been taken but 4 or 5 at a time are the general numbers. He was very anxious to know about the size and expense of our ships of War and Commerce, how they were manned, the length of their voyages and in fact every thing he had ever heard mentioned he wished now to know the truth of. He pressed us to stay with him longer but Williamson persisted in leaving tonight as he is much wanted at home in Mr Brooke's absence. We however promised to call before our departure returning to the boats. Having been with the chief about 2 hours I was very much pleased with this man's intelligence and anxiety for information and have not slightest doubt that if Mr Brooke and he knew each other the good feeling which now exists would be strengthened but on account of his leg he cannot leave his own Country. Mr Brooke at present has too much to do to think of visiting Serikei. He has no legal child to succeed him and those of his relations who are there are either too little liked or possessed of too little talent to become the head of this people. Should his death happen soon before any one springs up capable of succeeding him, this large town will be broken up and the probability is that two thirds of the population will settle at Sarawak. At about 6 pm we went to take leave of Patingi Abdul Rahman and dropped down with the ebb tide, the Dattoo having previously sent about seven fowls and a goat as presents during the night. We anchored in the Rejang on acct. of the tide and with the next

71

ebb reached the fishing village of Rejang. Here we went ashore and were perfectly astonished at the height of the houses; they are in some instances more than 30 feet above the ground and one could not have been less than 40 feet. You ascend them by ladders which can be raised when an enemy attacks them, when from above a shower of spears descends on the assailants who work hard with their axes to cut down the posts, covered from the spears by their large shields being slung over their backs. I should fancy it is but seldom they succeed in thus throwing down a village as they must first cut perhaps two or three hundred posts of the hardest wood under a shower of darts from above. The enemies they fear are the Serebas Dyaks who, together with the Sakarran tribes, on account of their piracies by sea and land are the terror of the whole coast from Saribas Eastward and Northward to Tanjong Barram. The Rejang villages were at this time full of inhabitants who were employed fishing and apparently with much success. The fish they catch are dried and sent to Serikei in part for the supply of its population and partly for export. Shrimps form one of the principal branches of the fishery; these are skinned and dried in the sun where they form a portion of their food of which the Malays are very fond. They are to Europeans by no means unpalatable in that state but when cooked a little so as to be perfectly hot or when made into curries they are delicious. This being the only tribe of Melanoans I have had an opportunity of visiting, I was particular in observing such of their women as I had an opportunity of seeing, they being famed as the beauties of the east as are the Devonshire and Lancashire Ladies in England. Neither Williamson nor myself thought this distinction was unmerited amongst them. Those I saw were exceedingly fair and their features were much more decided and distinct than Malay or the generality of the Dyak women. I have amongst the Dyak tribes seen women equally fair and beautiful with any I saw at the Rejang village but such were perhaps more numerous in the latter place than amongst the other tribes. I can scarcely tell what to think of my taste now or how to comment, the results of my observation on a point so delicate to [prefer]. When I first came here I used to look upon the native women with disgust; now I can as easily discriminate the degrees of beauty as one resident in an European Country would there. Instead of saying degrees of beauty I ought perhaps to have said plainness or rather ugliness for certainly they are not a comely race, but as I said before my ideas from constantly seeing them have become so vitiated that what we call a pretty woman we look upon with as much pleasure or nearly so as we used at the divine forms at home. It is certain however that the Malays give the preference to Melanoan women and a young slave girl of that race will fetch a very much higher price than from any other tribe. The Melanoans have the advantage of the Dyak women that whereas by the scantiness of the clothing of the latter their imperfections of figure are all before the eye, in the former they are concealed (at least such was the case with this tribe) by their long Malayan dresses. I had my gun with me ashore which rather astonished the Rejangs [as] they had not seen one with two barrels before, nor had they ever before seen birds shot flying so that two or three swallow which I killed for their amusement very much pleased them. I also shot a large squirrel which was jumping from post to post of the houses and might from [word missing] height on them he could attain have fancied himself in his own jungle. After having staid ashore about 2 hours we returned on board and were detained an hour longer taking in Dried fish which the inhabitants brought off in abundance as presents to Williamson whom, though they had never seen before, they all new [sic] by name as do I believe every tribe on the coast. We got under weigh but after sailing two or three hours we brought up again as the natives would not go to sea till they had another tide. We humoured them and waited. I walked ashore with my gun but could not get a shot in the Jungles; returned to the beach which is of hard sand and extends from were we lay to Tanjong Jerigi, a point of land covered with Orran Trees and the eastern point and the mouth of the River (the opposite one is named Tanjong Lallang), a distance of about 4 miles. Here I found lots of flowers of 3 or 4 pretty species; Curlews and sandpipers they were all together in large flocks and in to which, as my business was to get them for food, I fired and with the two barrels generally bagged a couple of dozen. Sherip Husein who did the same for his boat was equally successful. As soon as we had fired the men came in and killed the birds by cutting their throats (although most of them were already dead), a Mussulman [Muslim] not being allowed to eat anything from which the blood does not flow. During the night the people got under weigh and having sailed a little distance soon were aground so that I think it probable a bar extends across the mouth inside of the two points. Being afloat again we pushed on and our people saying there was too much sea outside put into a small outlet of the Rejang which disemborgues [sic] at Tanjong Lallang. The next day was cloudy and threatening but as we had entered this place against our will so we determineed to get out at all risks, much against the wish of the natives who protested that the boat would swamp in the surf which was beating over the bar at the mouth of the River. We however pushed through, though not without very considerable difficulty. The other being smaller would not attempt it and found a passage where the water was less rough close in shore. By the assistance of a very firm breeze we reached the Batang Lupar River the same night and anchored in it about 5 miles from the mouth. The next flood took us up to Lingah which is situated on the right bank of a River of that name, a branch of the Batang Lupar. I only went ashore to the Court House, a tumble down place, and the whole Malay village seemed equally miserable. It is the residence of a powerful Dyak tribe whose houses I did not visit as our stay was short. They, though situated so near the great outlet for the Sakarran tribes (the Batang Lupar), have always been able to defend themselves from their encroachments and have frequently attacked the pirate boats though not precisely under Mr Brooke's government. This place is entirely under his influence and the three chiefs the Indra-Lelah, Lelah-palawan, Lelah-Wangsa were appointed by him with the sanction of the Rajah. I saw a great many fruit trees about the village; they export Birds Nests, Wax, padi and Rice. Williamson having settled the business on which he was sent, we left with the ebb tide and anchored again in a creek of the Batang Lupar where one of our men who had been sick all our voyage died. We tore off part of one of the sails and wrapped him in it and made all sail for our own river which we reached on the following morning and pulling up against the tide were home in time for dinner and very glad we were having been out of cigars and every thing in way of drink for two days with nothing to eat but rice and fowls.

May 9th. Left Sarawak on this day with Mr Brooke, Captn. Bethune RN and Mr Williams, the Geologist sent out by Government, together with Williamson on HMS Steamship Phlegethon for Borneo [Brunei]. The Schooner dropped down on her voyage to Singapore on the same day.

We arrived safely in Borneo after a voyage of 5 days which length of time was occasioned by fuel falling short off Tanjong Barram. We were also hurled amongst the shoals off the mouth of the River and in its entrance we struck on a rock part of a bar which is thrown across the mouth to prevent the entrance of shipping, a very narrow channel being alone left. We again got aground in the reach below the town on a bank of soft mud; here we lay in sight of the houses about 18 hours while here Pangiran Beder-ed-din [Bedrudeen], one of the greatest Lords of Borneo, came on board to visit Mr Brooke. When we reached the mouth of the River he was laying there with five boats on the water for some Laboan pirates who had been lurking about the Island of Labuh-an. He had then visited Mr Brooke who is his most intimate friend and protector as will hereafter appear. He is a brother of the Rajah Muda Hassim and a man of intellectual features and most dignified carriage. His behaviour perfectly astonished me, his natural good breeding and perfect ease of manner differing so widely from the lower grades of Malays whom I have hitherto principally seen. It is true he has been much with Europeans, having lived at Sarawak during all the time of Mr Brooke's stay there till within the last 12 months. He is described to me by Mr Brooke as a man of the highest order of talent and as [h]is brother is heir apparent to the throne he will doubtless become the real ruler of the Country. He is well disposed to Europeans and it is in a great measure on acct. of Mr Brooke that he has attained his present position.

On our arrival at the town we saluted the Sultan with 21 guns and the same number was returned from the Sultan's Battery and the Rajah's House. Mr Brooke invited me to accompany himself and Captn. Bethune ashore to wait on the Sultan. His Palace is a short distance up a River which branches from the Borneo River to the right; it is called the 'Sunghei Kadyan'. Having arrived off the steps of the Council Chamber, we ascended and seated ourselves on chairs placed for us on either side of the throne, Mr Brooke being on the right of the Sultan and Captn. Bethune on the left. The Rajah Muda sat between Mr Brooke and his Highness also on a chair in the body of the Room, it being divided in three partitions by the rows of posts which support the Roof, was occupied by the Pangiran. In the two aisles were seated the inhabitants of the better class. The Sultan came in after we were seated preceded by the State Sword and Spear which were both profusely ornamented with gold and the latter further adorned with hair and something resembling precious stones, but perhaps it was but glass as I could not narrowly inspect it. The Sultan himself was plainly dressed in silk, wide trousers with a broad cloth jacket lined with satin and embroidered with gold. His sarong was the black kind manufactured by the Illanons; his creese [kris] and handkerchief for his head were both adorned with gold as were those of the other Pangirans. His Jacket being open, his breast was exposed nor did I see one of the Pangirans of Borneo whose breast was covered. The throne was nothing but an old China bedstead of the better class, covered, gilded and ornamented with cloth. Before it was a footstool and rich carpet; beautiful mats were spread over and by custom the sultan ought to have sat in it crossed legged on the mats but he merely used it as a chair. The conversation during this short audience was of the most trivial kind and at parting the Sultan shook us all by the Hand as did the principal of the Nobles. The Present Sultan on account of the misrule of his predecessor for the last forty or fifty years does not possess in reality one half of the possessions anciently belonging to this Kingdom which was one of the most powerful of [the] Malay States. He does not even possess the title proper to his rank. When [Captain Thomas] Forrest was there in 1776 the chief of the kingdom was styled 'Yang de pu-Tuan', meaning the Lord who rules. The first minister next in Rank to the Yang de Putuan held the title of Sultan; then came the Pangirans who were each Rajahs in their own provinces and at that time are stated to have been but about 15 in number. This I think must have been a mistake as they are now probably 500 and as no new ones can be created the only [means] of accounting for such a number is by the increase of children. But this increase is impossible, all of whom whether Legitimate or otherwise are styled Pangirans. The Rajah Mooda told Mr Brooke that at the installation of the present sultan the Country could not afford to create him Yang de Pertuan on acct. of the expense of the Ceremony, 2000 £ being scarcely sufficient to defray it. Every house belonging to the nobles is covered with cloth of gold, their wharfs, posts, boats etc. with satins. The palace being [illeg.] belonging to the regal dignity must be profusely ornamented so that as I said before the country having been impoverished by the rapacity of many Tyrannical Sultans who were frequently murdered by Pangirans could not go to the expense necessary. I think it probable that the next sultan will resume the ancient title.

The present Sultan is certainly a fool, probably something of a rogue, and on the whole has been pronounced nearly an idiot. He is no less feeble in body than in mind as although his seraglio is filled with young women he is incapable of begetting offspring. Two or three children born by his women are the unbeknown offspring of some of the Pangirans but the old Sultan is very proud of them and tries to cram the people with the belief that they are his own. On being told that one of them very much resembled a Pangiran who was known to be his father, the Sultan accounted for it by saying that when the mother was pregnant she thought much of this Pangiran, hence said he the resemblance. On his right hand he has two thumbs, the one of which springs from the lower joint of the natural one and turns in-wards to the hand. In this however he is surpassed by one of the nobles who has an extra thumb to each hand. Had Law been strictly attended at his elevation to the throne he could not have become Sultan as it is expressly forbidden that any person having the slightest blemish should be put in possession of that dignity. In the evening I again accompanied Mr Brooke to [illeg.] to pay our respects to the Rajah Muda Hassim, the Heir apparent. He appears to be a man of amiable qualities and features which betoken benevolence but at the same time they express but little determination of character. I am told however by Mr Brooke that in an emergency the Rajah shows himself a man both of determination and courage. He is very short in stature, probably not more than 5 feet 1 inch. With the exception of Pangiran Beder-ed-din he is the only one of the nobles whom I saw wearing shoes. The sultan is his nephew though much older than himself. At present the Sultan and Muda Hassim are not very friendly, another Pangiran (Usup) having usurped the Rajah's power in the state during the absence of the latter at Sarawak while attempting to quell the rebellion, a period of I believe 6 years. During this time Usuf or Usup cultivated the friendship of the Sultan and people and raised a party in the State which appears likely to give some trouble before Rajah Muda can attain his own. Usup has moreover declared to Mr Brooke his designs upon the throne, having said once in conversation that when the present Sultan died no one should succeed but himself. This, however, Mr Brooke is determined to prevent and [as] the Pangiran Usup is rich and powerful he will certainly bring ruin on himself. He is described as a person of high talent and is thus the more likely to be troublesome. Unless the Sultan forsakes the party of this man, they will probably both be involved in the same ruin as Mr Brooke is determined to rid the Court of Usuf and his adherents and to place Mooda Hassim and Beder-ed-din in power. If the Sultan raises any objections he will probably be dethroned but this he is not likely to do as he told Williamson that provided they would give him women and money and let him live quietly in his palace he cared not who ruled nor if the Country went to the Devil. He offered Mr Brooke the whole Country from Tanjong Dattoo to Tanjong Barram, an extent of nearly 200 miles in length including rich Countries and powerful tribes. To this was to be added the title of Sultan but Mr Brooke would not consent to such a dismemberment of territory, having the interests of his friend the Rajah at heart at the time more than his own. This seems indeed to be frequently the case with Mr Brooke. I never met any person nor do I believe another such exists who with the same means would care so little for his own interests provided the setting them aside would favour his friends or benefit his race. The principal object of this expedition was to give Mr Williams the Geologist an opportunity of examining the Coal seams which run all over the Country. That which was most easily accessible was exposed in part in the bed of the Kiangi River, a small stream up which the smallest boats cannot pass. Williams gave it as his opinion that this coal of excellent quality may be worked for 10 years and produce 100 tons per day. For that space of time, on account of the jealousies between Usuf and the Rajah, it was very difficult to get at the Coal as when the former heard of a seam being discovered he sent orders to the person to convey himself away till we were gone. By this practice we were often baffled. The Kiangi river is situated at the back of the range of Hills which extends from the mouth to the town parallel with the River and in fact forming its banks as the valley through which it runs is of only sufficient space to allow of the passage of the waters. Where the footpath crosses the range it is about 300 feet high but in many places they rise to 700 and 800 feet. The river is about a quarter of a mile distant from the main river and the road leading to it is called the Jalan Suback. On the left of this road is the burial Ground and in the middle of it stands a fine specimen of the Upas tree; it is about 40 feet high without a branch. At this height it throws out wide spreading branches which gradually towards their extremity incline towards the ground and give the tree a very graceful appearance. On being cut with parang the sap flows freely from it so that a considerable quantity may be collected in a very short time. At this time it was neither in flower or fruit so that had I descriptions (which I have not) of the Javanese plant I could not have determined whether this was the same. It is called 'Bina' by the natives, the Upas being a climbing plant a species of Strychnos I am told which grows along the Coast to the Westward; this plant I have not seen. The sap of the two is mixed together to prepare the poison used by the Dyaks; it is not however very immediate in its effects as several of our people were wounded with the Sumpit [blow-pipe] arrows at Sakarran without any bad consequences ensuing. The common people tell just such tales of this tree as did those of Java of the tree 'Ipoh' or 'Upas', but no one need be now told that they have not the slightest foundation in truth. The tree is remarkably scarce in this part of the Island, this being the only one I have heard of. It is not impossible, however, that they may be more plentiful in the Jungle which Europeans have not yet penetrated. During my stay I did not once set my foot in the old Jungle because when Mr Brooke asked permission for me the Pangirans advised me not to go and insisted that if I when [sic] it must be at my own risk because Usuf's party would not scruple to cut off an unprotected European in the idea that such an act would raise him in Usuf's estimation. Borneo town has been so often described that as I can add little or nothing to that which is already known I should pass it over altogether. But if one of my friends for whom this journal is intended should not have seen or shd. have forgotten the acct. of it I will just say that the houses are built as other Malay houses on posts, but these are placed on Mud Banks in the River which is very wide at this place. The houses are placed with more regularity, being divided into solid Squares and Lanes through which the River flows and gives them the appearance of what indeed they actually are: Water Streets. The grand street is formed by the Main River [and] on it are situated the houses of the principal Nobles. The Sultan's Palace is as I said before on a branch of the River to the right. It is surrounded by a high Pagar [fence] of close rattan work so that when near you can see nothing of it but from the ship you could discern a substantial and large house surrounded by smaller buildings. I never heard of any English having been allowed to enter it, the Sultan's private reception room as well as the Council Hall being outside the Pagar. Every thing about it seemed neat and nicely arranged and [illeg.] —ly mounted lelahs [small cannon] which command the landing place had their carriages prettily carved. To the right of the 'Astana' or Palace, that is to say on the side nearest the town, were placed many fine old guns of large calibre; I am told that they were principally presents from the Spaniards. In their present state, being without carriages and overgrown with grass, they are of no service. Were the Mosque knocked down which occupies the adjacent point of Land and a battery erected which would in a great measure command the whole town would render these guns of some service to the ruling power. The point is however very low and perhaps would be found scarcely available on that acct. Pangiran Usuf's House with the exception of the Sultan's is the largest in the town. It occupies the foot a steep hill about 150 or 200 ft high which is generally fortified by who ever can first seize upon it. During the Civil Wars Usuf has wood cut ready to fortify it now but though he is anxious to do it he cannot as [he] well knows such an act will be prevented by the opposition and should he still insist they instantly resort to force which he having the power would repel them. All the Rajah and Beder-ed-din can do is to defend their lives till Mr Brooke's arrival with the European force he has promised to bring to their assistance. The power of Usuf by his intrigues and the religious differences of the Chiefs is greater than that of the Rajah. The exports of Borneo are pepper, gold and camphor of which together with Malacca canes and Rattans large quantities are annually sent to Singapore both in native praus and square rigged ships. The East India Company had once a factory here which was abandoned when the disturbances respecting the Government broke out. An American ship of War was here a month or two before our arrival. They anchored outside the River and sent up a force of 80 armed men under the command of the first Lieut. to tell the sultan that they wished to make an exclusive commercial treaty with the Country at the same time they requested the sultan to go down to the Ship to wait upon the Captn. who was laid up with the gout. They also wished to be supplied with a hundred dozens of fowls and 100 oxen, the latter at the rate of 4 per tail. This demand was no doubt made to ascertain the resources of the town as a place of refreshment. The Captn. must have been strangely ignorant of Malay manners to request the sultan to wait on him; he must, too, have had a tolerable idea of his own dignity. The answer they gave to the first question was that they could give them no priority in trade as negotiations with Mr Brooke and the British Government were on the table at the time. They were then very curious to know how Mr Brooke had acquired his influence over them and at last requested one of the Pangirans to go down and see the ship. Beder-ed-din offered to go in two days but they sailed before that time being elapsed.

After about a fortnight's stay in Borneo we crossed over to Labuan, the Island laying off the mouth of the Borneo River on which the Government has an idea of making a settlement. It was for the purpose of ascertaining whether this place is available for such a purpose that Captn. Bethune was sent out. We remained here a week, the Schooner being anchored in nine fathoms within 300 yards of the beach so that the wood was easily taken in. This harbour very much resembled that of Singapore and should a town be built the same arrangements will suit admirably for this place that were used at Singapore. Labuan Harbour is sheltered by Islands and reefs of rocks mostly above water which seem to break the force of the sea but as on the North and North East which are the only points from which heavy gales blow here it is well sheltered by the Island even without the reefs and Islands as well as the main land of Borneo which is distant about 20 miles it would be perfectly secure. On the right hand bank of the River is a fine beach which extends as far as the Arab's grave a distance of 3 miles by the beach. That part nearest the River is Jungle and in it as in the Jungle elsewhere through the Island grows a magnificent species of Areca. It is called by the Natives Nibong but it is different altogether from the Nibong of the Sarawak River, being amply furnished with spines on the stem which is double the height of the common species, reaching at the least to 80 feet and in many instances more the leaves as well as the stem are more and longer and graceful than that of the common species which has a clumsy appearance when placed besides its con [illeg.] this species and the Betel Nut. I found here a species of Podocarpus in the Jungle forming a tolerable tree but it was not in fruit. The Arau tree

(Casuarina littoria) lined the Beach and the magnificent Orchideous plant Calanthe veratrifolia was every [where] displaying its large heads of snow white flowers from the rotten trunk of a tree or the decayed vegetable matter on the ground. Other Orchidaceous plants are very rare here but I think that if I had gone up the little river which flowed into the Harbour opposite our anchorage I should have found some as none of the places I saw were at all likely to produce them, they being generally found in the greatest quantities on the trees overhanging a river or creek the vapours exhaled from which are necessary for their support. The second division of the distance between the River and the Arab's Grave is occupied by a large plain about the size and exactly resembling that of Singapore. The sea has at no distant period covered this place and it would be one of the finest spots for Cocoa Nuts I ever saw in my life but it will be wanted for Villa Residences for the Merchants and others. Very excellent fresh water was found by Captn. Bethune by digging about 3 feet deep and that with[in] a hundred yard[s] of the beach. The last division of the distance is again Jungle, full of low palms of small species. There is also here a large pond of fresh water which in some places approached the beach to with[in] 10 yards. It appeared shallow and I thought that it might perhaps be dry at some seasons, but I was soon undeceived for having shot a bird about the size of a thrush from one of the trees overhanging the water, it fell into the pond and before my boy could reach it a large fish rose to the surface and carried off my bird so that the pond must be supplied by springs. The sun would soon evaporate the water although it is well sheltered by trees on every side.

During our stay Mr Brooke and Captn. Bethune wished Williams the Geologist to see a small river of coal on the North East part of the Island. I obtained permission to accompany [them] and in this place it is. I must note Captn. Bethune's kindness to me. He never went surveying to any place where plants were growing without asking me if I should like to go and if I thought that any spot in the neighbourhood promised better than another or than that to which his duties called him. Another boat was ordered to carry me to it. His urbanity of manners endeared him to us all and without about two exceptions I have never met a man of such gentlemanly manners and so kind a heart ardent in the pursuit of his duty. Day after day where ever we were we saw him out with boats fixing the points and laying down channels, dangers etc. Such a man ought surely to be advanced in his profession or have some post of trust and honour confided to his guidance. When he can speak the language a little better and has seen something more of the habits of the Natives, I think Captn. Bethune would make a good governor for any settlement in the Malay Countries, he being one of the few men I have seen whose temper would lend itself to this disposition. Mr Brooke is the only person I believe in the whole world who is born for a ruler of Malays and for that purpose and the amelioration of the condition of the Dyaks I am persuaded he was ordained, so perfectly is he fitted for the task. May his power and influence continue to increase for many years. But 'revenons a nos moutons'. I accompanied the expedition to the Coal river and in going we had a good view of the mountain Kinaballoo [Kinabalu] though we were forty leagues from it. At the part of the Island where the small coal river is situated there are some very magnificent trees. I measured one which had fallen; its trunk to the branches was 120 feet in length. And I saw a large Dammar which would have measured more. Williams having found that the coal was in too small quantities to render it worth working we returned in the evening to the Ship. On our way from Labuan to Sarawak, having fallen short of wood, we entered Palu, one of the mouths of the Rejang. There were about 20 families of Milanones [Melanaus] living here and as many on the other side towards Tanjong Serie. With their assistance we soon completed our [word missing]. As we had anchored close to where it grew I found here a kind of Banyan tree, a ficus, each tree covered a great space; it was growing close to the sea shore.

October 24th 1845. Left Sarawak with a fleet of eleven boats under the command of Mr Williamson for Sakarran River Mr Brooke having received intelligence by Lingi one of the well disposed chiefs from Abang Kapi the principal Malay in the upper part of the River that a fleet of 18 boats under the orders of 3 Sereibs Amal Lingir and Bakar, was going out piratting [sic]. We were sent with orders to kill the Sereibs and disperse the boats and in the third day we arrived at Lingah River and anchored at the mouth of it. Williamson went up on the following morning to learn the news and get a force of Balow [Balau] Dyaks to act in conjunction with us. When he returned he brought news that two days since the fleet had been attacked by the Balows at a small

River inside Serissan, the high Island-like point at the Mouth of the Batang Lupar. When they saw the Balows who were firing guns and making all possible noise they jumped into the water leaving their boats and swam for the Jungle. The Ballows [as] soon as the[y] could did the same in pursuit but they got but one head, this was that of the Dyak Chief Api Biagi. All the boats together with everything they contained became the property of the Dyaks of Banting and, [being] all amongst the plunder was found the gold ornaments of the Sereibs [sharif] which it is probable they will have some trouble to replace. The fleet of boats prove of use to the Dyaks as they are all new and the Balow boats are rotten. We ascended the River to Pa Mutas where we stayed in the deep water for the night as it was spring tides the bore may be expected. This is the site of the town destroyed by the Honble Captn. Keppel of the Dido last year about this time. It was built by Sereib Saib [Sahap] and his followers on their retreat from Sadong and was not yet finished when attached by the crews of the Dido and Phlegethon. After being obliged to run and having lost everything belonging to him it is said that this Tyrant of the Coast died of a broken heart in the interior of the River. His associates Sereib Mullan, Dattoo Maharaja, the Sereibs we came against and others, ran in different directions for the[ir] lives and the place was burnt. The posts were still standing at our visit. The bore came up during the night with a great noise but we being in deep water it did no harm to us except of making the boats pitch heavily. As soon as the bore had passed on the following morning the fleet got under weigh for Utang Kapi's village where it arrived after about 3 hours hard pulling. This is situated just inside the Sakarran River at its junction with the Batang Lupar which comes down from the Lakes and the Country surrounding them. Our force of Balows' thirty two Boats were with difficulty kept from pushing ahead, which would have endangered the lives of any poor unfortunates they might have met. In their ascent they have scoured the Jungle for stragglers from the conquered fleet and have caught four heads, one man escaping them. The object of coming here was to explain the intentions of Mr Brooke to the well disposed portion of the people. Abang Kapi is as I said before the principal Malay and Mr Brooke places trust in him. He has been ordered to build a fort which Mr Brooke is to supply with guns. It is to command the river and so prevent the egress of head cruising fleets.

Gasing, the head chief of all the Sakarran tribes, is the friend of Kapi and wishes to give up head hunting and to practise trade instead. He is a very fine looking elderly man and dresses very neatly in the Malay fashion. Having heard that the three Sereibs were in the interior Country at their houses and days after our arrival, Gasing, Bulah, Lingir, Sahban and others of the principal chiefs were called and the necessity of cutting off the Sereibs referred to them. They all agreed to go and surround the House, Kill the Sereibs and bring their heads. When asked whether his people would follow he said he would kill those who dared to hold back and that his children had burnt houses and killed people enough in their time so care very little about taking a few heads more although they were Sereibs. They went up that night to collect their people. During the time we were waiting their return the news came of the death of Gila the chief of Gila Beranhi [illeg.] the fiercest of the chiefs who still wished to continue their piratical practices. Now that Biagi the chief of Pelissow and this man are dead I think they will pirate no more. He died of a short illness (but 2 days) and it is by the natives ascribed to Mr Brooke's charms. On the fourth day since they left us the Dyak chiefs returned unsuccessful. The Sereibs had been there but at [the] noise of our guns which were fired to call our chiefs together they with the assistance of the Kangluna Rajah of Seribas conveyed their wives and children overland to that river and followed themselves. Sereib Amal is the son-in-law of Panglima Rajah. Gasing offered to take his men over to Seribas and [illeg.] them or take them by force if Williamson wished it. This could only be done by making the powerful and warlike tribes of Seribas the enemies of the Sakarrans and it was not insisted upon. We further heard from Gasing that when Gila was dying he called his tribe around him and requested them all to follow the advice of Gasing for said he if I had not opposed his and Mr Brooke's counsel by sending my children under Api Biagi I should not be now so near my end. One of his, Gila's, near relations told Gasing that if he was not allowed to revenge the death of his friend Biagi on the Balows he would go and build his boats at Seribas, which river he would descent and attack the Balows from the Sea. Gasing said in answer that he and his tribe were nearer than the Balows by that route and it [illeg.] thing who are attacked for all would become his enemies and soon destroy him. Tomorrow all the Chiefs are going to swear according to their custom by killing a pig that they will be the Tuan Besar's faithful friends. At the mouth of the Lingah on our return the Orang Kaya of Lundup who accompanied us obtained one of the heads from the Lingah Dyaks as his share. He founded his claim on its being wrong for the Balows to attack the fleet because, said the old fellow, we were sent and you were not. The head [had] been smoked and the features were black. The lips were drawn backwards so as to expose the teeth, which were filed to sharp points. Two days after this we all arrived safely at our own houses in the beginning of Nov. A few days afterwards I found in the Jungle one of the finest climbing plants I ever saw. It was covering a high tree with one mass of crimson. I know nothing of its name, tribe or order. On the Sunday after our return I found a Rhododendron with yellow flowers epiphytic on the stem of another tree. The bunch is magnificently large and the leaf large and smooth, habit dwarf and drooping. From this day to the 20th I was employed searching Sungei Bidit for more of the Rhododendron and having found two both epiphytic on the main stem of other trees.

[Nov.] 20th. Found a yellow Rhododendron this morning. Wrote a letter home in which I named the Rhod. Brookeanum in compliment to James Brooke Esq. of this place whose kindness to every person about him deserves a more lasting monument than a flower can bestow. Made preparations for the ascent of the River to-morrow.

Nov 21. Left Sarawak with the flood tide and stopped for the night at Seniawan. Slept in the boat and was dreadfully pestered with mosquitoes. Also a little in dread of the alligators as I saw the body or rather bones of a man taken out of one which had been caught three days since. The man had been carried out of his boat by the beast whose length was 15 1/2 feet. It was a male.

Nov 22. Left Seniawan at 8 am and pulled against a strong current to Pincallum Bos [Pengkalan Bau] where gold is extensively worked and which we reached at half past one pm. We soon after passed the villa of Subah belonging to the Sauh Dyaks and two hours afterwards reached Lubong Angin where we intended to pass the night on the [illeg.]. Above the cave at the height of sixty feet in a place apparently inaccessible I saw several plants of the Rhod. Brookeanum and was of course anxious to get them. The men all said it was impossible. I offered a dollar a plant for all they could get in good health and left them [to] consider the proposal while I trimmed my lamp to prepare for the entrance of the 'Wind Cave'. I had been told that no Malay had ever entered it but that the Dyaks frequented it for birds nest. I was also told that a stream of water ran through it and wind never ceased to blow in it and also that to walk from one end to the other would take three hours. I entered this evening merely to have looked into the mouth of it and penetrate it further tomorrow. I commenced by following the stream which was flowing out of it. It was knee deep with [illeg.] and water and had worn itself a course amongst the rocks. The faint light of my lamp enabled me to see that one part of the cave was deeper than the other, that which the stream followed of course being the deepest. I ascended the highest part which appeared to be composed of broken and rotten stalagmites though there was no appearance of them ever having been attached to the roof which was water worn, arched and marked as if water had flown through it while the stone was in a soft state. I found that the stream did not rise far in the interior but was formed by water dripping from the roof. The flooring to a little way in was much higher than at the entrance and there I heard a noise as if a gale of wind were blowing in a vault and had no doubt of the propriety of the cave's name but I had not advanced fifty [? feet] further before I saw light at the other end and myriads of largish bats fluttering about the roof and crowding to two openings; the larger one was straight before me, the other up a corridor to the right. I advanced up that to my right, followed at some distance by the two men who entered with me and who when they saw the light had lost little of their excessive fear of darkness. The light proceeded from a little dome like the steeple of a church up which was not difficult to climb with the assistance of the broken Stalactites. I then returned and followed the other opening which was more considerable and the main opening of the Cavern on that side with some difficulty on acct. of my feet being naked and the excessive sharpness of the tops of some stalagmites piled at the opening. I got outside but could not prevail on the men to follow. I had left them in the other corridor. I was the more anxious to get out as I thought I should find land shells having seen a great many dead ones which had evidently been carried to the bottom of the hole from outside by water. I found four small spiny shells which I put in spirits. While looking for them I found amongst the Rocks and dead leaves a Begonia of no beauty a small worthless Justicia and two plants of a beautiful Anoectocheilus the leaves are like [? Aeschynanthus] but not so strongly marked with gold perhaps because it is that species which I believe comes from Ceylon. I returned through the Cave passing over immense heaps of what the natives told me was the Bats' excrement; it had a nasty smell at least. The cave had probably from this cause; it might prove to be a kind of Guano. The Wind or rather sound of wind which gives name to the place is caused by the fluttering of the wings of the innumerable Bats. The whole extent is probably 200 yards in length and 20 to 30 yards in breadth inside direction S S E in unstratified blue Limestone through which run veins of Quartz. On my arrival outside I found myself indebted for 3 plants of the Rhododendron which Si Tekei had obtained by using the [illeg.] climbing plants for support during [illeg.]. The Butterflies had been very numerous about the fragrant shrubs on the Banks. We passed several of the Gigantic Japangs, three plants of the magnificent Clerodendron bethuneana. At night we slept opposite Lubong Angin and were free from the attacks of mosquitos and sandflies.

Nov 23. Left Lubong Angin at 8 am intending [to] call again on our return to look for more Rhododendrons and Anoectocheilus. In three quarters of an hour we passed the Sauh village, Incotong or Passir Bruang. Some of the Dyaks were standing on the Bridge which crosses the River here and we promised to call as we came back. Soon after I observed an Orchideous plant on the top of a high tree and sent a man after it. It proved to be an immense mass of Vanda like plant which I have previously sent home though I have not seen the flower. On this tree were about 100 drooping spikes of flower each from 9 to 10 feet long; they arise from amongst the leaves which clasp the stem at their base, generally one raceme of flowers to each stem but very frequently more. The flower is more than 3 inches in diameter, light sulphur yellow spotted and blotched very much with rich cinnamon brown. The labellum is purple in the centre and spotted with brown towards the edges. The flowers as they get old do not fall off for a very long time but gradually lose their rich blotches and acquire their yellow color so that the first flower on the stem by the time the others at the end have opened is of the richest golden yellow. The flower stem is rough and the petioles of the flowers are protected by a sheath above their upper surface as long as themselves. They are covered thickly with a brown substance resembling the moss of the Rose. The leaves are of a light green colour and leathery texture, a little more stiff than in Aerides odorota of the same shape 18 in. to 2 feet long. The roots do not arise amongst the leaves but are wholly confined to the bottom of the stems where the leaves have long since fallen. They are large but not so thick as might have been expected from the size of the plant; the stems are 2 to 6 feet long and grow horizontally from the tree. The scythes flowers hang down beneath. Having been delayed 1/2 an hour to get this plant we pushed on again and at half past ten came to a place called Battu Bidi on the left side of the River. Here were 2 or 3 miserable Malay habitations, the occupants of which live by bringing antimony on the Sarawak. It is found a little way in from the banks of the River but is better quality as [sic] that from Seniawan. At 11 1/2 h am we passed Sunghie [Sungei] Jagusi, a considerable stream inland of which is a tribe of Dyaks of the same name. Just before reaching it I saw two horned owls but could not get a shot at them. With a violet colored kingfisher I was more successful. At 12 3/4 pm [saw] in passing a few fossils similar to those at Battu up the left hand River. They were in Limestone disposed in layers at an angle of 48°. At 2 1/2 pm passed Sunghie Kirit on the left and soon after passed another fine bridge (there being one opposite Kirit) in the neighbourhood of which on the right bank is the village of Bokong inhabited by Sauh Dyaks. It contains 10 or 12 Lawangs. The Country we have passed to day for more than half the distance rests on Limestone which has been worn by the water so as in many places to over hang the river 15 feet. The latter part was a kind of sandstone not much above the waters in general but in some places 20 or 30 feet. Here the banks are high and well wooded with beautiful trees which were [sic] we lay for the night met overhead. We slept a little above the last village.

Nov 24. Started this morning soon after 7 H and at 8 passed a small stream on the right called Serekin and Lubok Pilin, a wide part of the river immediately after Sunghie and Tubah. On the left was the next River. This we passed at 9 H and half an hour afterwards a river on the right of which I could not learn the name here I gathered a pretty climbing plant with bunches of white flowers in the axils of the leaves in the way of Stephanotis floribunda but they are without perfume. At

10 we passed [a] bed of large pebbles called Battu Kaladi. At 11 I found a pretty and very minute Arum growing under the waters of the River in shoal places. The flowers were long narrow colourless and almost without stem; the leaves were curved, short, broad and without petioles growing close to the earth at the bottom of the River. It is very pretty and neat. At 11 1/2 H passed Sunghie Candong on the left and 3/4 of an hour after S[ungec] Kasong on the right. A little further on I saw a plant of the very magnificent long scarlet flowered Aeschynanthus, two plants of which I have before discovered but was not able to preserve. Wild Serih (Piper) was growing on the opposite Bank. In a tree near the last I gathered a very curious flower which puts one in mind of the stinking Stapelias. It is an oblong fleshly [illeg.] climber epiphytal in the hole of a tree full of the richest vegetable matter (and the most sharp biting ants I ever saw; they drew blood at every wound). The flowers are in bunches from the axils of the leaves, having an uncommon stem about 6 inches long from which the others about an inch and a half long spring forming an umbel which is pendant. The inside of the flowers is brownish purple but all the parts connected with fructification pure white which makes them very pretty. I took the plant but doubt whether I shall save it. Our journey today has been through high and well wooded banks of rich loam. At three o'clock we entered the Sunghie Sebuloh to the left. This stream is one third the size of the main River which is itself small and during today and a great part of yesterday we have had much difficulty to get the boat over the Shoals and fallen trees. Just inside the place of the Gumbang Dyaks where we slept for the night. Our journey since the night of the second has been beyond the furthest the Europeans have hitherto penetrated.

Nov 25. Started early for the village and found the road very bad. We crossed the Sebuloh several times and were afterwards told by the Dyaks that it had its rise from springs on Gunong Api, the mountain at the foot of which the village stands. We learn that we cannot possibly get up any further in the boat but that a long day's walk will bring us to Seringash, the mountain in the Sambas Territory where it has its rise. There these people have another village. We walked about 5 or 6 miles from our boat and as we approached the mountain the Jungle contained nothing but large fruit trees, principally Durians, Sebows, Tampuri, Lansat, Champadak, Sagos (in the low places) Mangosteens (Secup

Dyak) and as we approached villages Cocoa Nuts and Cinangs in great numbers. The perfume they all exhaled was most delightful. They are just ripening but the Durians and Tampuis alone were perfectly ripe. Of these we ate plentifully at the village. The hill on which they live at the foot of G. Api is plentifully supplied with water from many springs, all of which find their way to the Sebuleh at its base. The village contains about a hundred doors and [is] situated on the borders of the Sarawak territory so that part of their revenue goes to Sambas and part to Mr Brooke. The Orang Kayas were very much pleased to see a white man amongst them. Most of the people have never seen one before as they are not clever at managing boats and seldom come to Kuching. I went over the village with one of the Orang Kayas and found it more convenient and clean and the houses more commodious than any I have before seen. It is adorned with two head houses but they do not contain many of the disgusting trophies of their wars. Indeed, from all I could learn they appear to have been fortunate in a long state of peace. Their houses are all detached but the raised platforms are connected. They look very pretty amongst the fruit trees and enormous boulders of porphyry (Williams) which cover the face of the hill. The Orang Kaya explains the fact of different nations rejecting certain animals as food in a novel manner. He must have heard it from the Malays as I never heard of the father of our race being known to Dyaks. It was thus Adam by his wife Anal had many sons and he divided the whole animal world amongst them. To his first born whose descendants are the white he gave permission to eat of anything. To the father of the Malays he forbade pork, to the Dyaks Goats and Oxen and so on through the whole of them. Their God is as with others on this river named Juwata but they know nothing about him. They burn their dead as they say the person thus goes up to heaven in the flames but if they buried him he could not. They [know] nothing about the origin of the custom and when asked merely say that it is 'adat lama', a very old usage. I gave the head man some handkerchiefs as presents and every one begged me to stay two or three days to eat fruit with them but this I cannot do as I wish to get the plants I have collected safely into their glass cases. At the Orang Kaya's house I saw a long string of fine land shells but they were spoilt as specimens. In the evening the Chiefs and inhabitants came to the Head House, each bringing his present of Fruit Rice Eggs etc. in his [?hands]. In return I distributed tobacco amongst them.

Nov 26. Having refused all entreaties to stay on acct. of the plants in the boat, I left the village at 10 AM. 6 or 7 Dyaks were following with fruit and one of the Orang Kayas was with me. The other sent his son and begged to be excused from personal attendance on acct. of sickness. About one we left the landing place of these friendly Dyaks. I wished to stay longer but my plants were to [sic] valuable to be thrown away. I have almost promised to call at Jaguoi on my return. In our descent we took two plants from a large of [sic] Aeschynanthus I have mentioned it before. The flower including the exserted stamens is 5 inches long; excluding these 4 1/2 ins. The colour of the tube is yellow for one half its length, the limb and remainder orange scarlet inside. The lower and two side divisions of the corolla is a stripe of black, the same shape as the Division of the corolla in which it is found and situated about 1/3 of its breadth from the edges. This gives it a very beautiful appearance. The leaves are verticillate, 3 to 4 in a whorl, fleshy smooth oblong pointed with a short petiole dark green above light beneath, 4 to 5 in. long, 1 1/2 to 1 3/4 broad, epiphytal generally in the rifts of the birds nest Fern. I should ask Dr Lindley to name it Aucklandiae after Lord Auckland [illeg.] to have a new Fern as more appropriate and equally beautiful [illeg.].

At half past 3 PM we arrived at the Sauh village of Klakong where we passed the night. I find on examining [it] that it is composed of a single row of well built houses to the number of 30 or 40 instead of 10 as I was at first told. In the evening I made some enquiries respecting the Kyan people from a man in my service who has spent a great part of his life amongst them and visited their country at different times. I learn from him that they are a very numerous people, very much more so than the Dyaks. They live in houses built in one terrace as do the Dyaks but they are much more substantially built. The posts are strong and generally of Bilian or some hard wood. The roof is of Billian planks instead of thatch; the sides and flooring are also planks. They cultivate padi extensively and abound in pigs, goats and fowls. The men dress as do the Dyaks as far as their waist band is concerned but are extravagant in having it of very great length so as to go many times round the body. In peace they wear no other covering. In times of peace their ornaments are said to be Ivory amulets but it is probable they are but bone. The

Malays sell them at a high price. In each [of] their ears at the top they wear a tiger's tooth, the point of which is directed forwards. In war they wear jackets made of the skins of goats bears etc.; on their heads a curious cap fashioned in front like the face of a man. Above it is surmounted with the feathers of the Argus Pheasant. Their arms are shields, swords (Ilang) convex and concave in the blade, the handles and sheaths of which are ornamented with human hair stained of different colors. This they buy; it is not the hair of enemies. Their spears have short broad points and long handles. They do not throw them. Their points are steel and are pointed to inflict fearful wounds. Their steel is collected in their own country as it appears from beds of metal or Iron stone. The women wear a cloth or rather two cloths from their navels to their ancles [sic]. The one covers one side and its two ends are tied on the opposite hip; the other covers the part left exposed by the first and is tied on the opposite side to it. These cloths are generally of dark blue with red borders. Their ears are pieced so that the lower part reaches to the shoulders. In this hole or slit they wear heavy copper ornaments. Their necks are ornamented with beads and their arms with Ivory. Amongst the men the use of ivory is confined to those of distinction. The women and men both wear long hair but that of the latter is cut straight across the front above the forehead as with the Dyaks. Every one, both Dyaks (their enemies) and Malays, allow them to be highly brave but afraid of firearms because not accustomed to them. They merely know their deadly effects. In their government they are similar to the Dyaks, each village having its chief who however bears no title but is always called by his own name. Their dead they place upright in a hole which they form in a large hard wooded tree for the purpose. Over his head they put a large stone to cover the hole. The tree is left growing. When their chiefs die they immolate human victims together with pigs, goats etc. The bodies of the human victims are placed in the tree with the body of the chief [and] the pigs are eaten. Their Country abounds in fruit and the Rhinocerus [sic] is found. They do not eat monkeys, snakes etc. as do the Dyaks. The inhabit the country interior of Banjan Pontianak (far inland), Coti Bruni, Barram and Serekei. The men also have their ears stretched as do the women and my informant says that women alone are tattoed {tattooed] and those only from the elbows to the fingers and from the thighs to below the knees. He says that is the Orang Bekatan [Bukitan] and Satoiw near Bintulu who are tattooed all over their bodies and that these are not Kyans but Dyaks. These are also very clever with the Sumpitan and he further says that the Kyans do not use this weapon, which is different from anything I have before heard as Mr Brooke supposes the Kyans to be very clever with the poisoned arrows of the blow pipe. The Kyans are supplied with salt cloth etc. by the Malays who get camphor in exchange on which they make large profits. The Kyans take the heads of their enemies and keep them in their houses like the Sakarran and Seribas tribes of Dyaks. They live as do the other wild tribes with but one wife. In a case of adultery the faulty man is killed by the injured husband. They hold it as an axiom that whatever is the temptation the man alone can be the guilty party.

Nov 27. Started at daylight and breakfasted at Incotong. This village contains 20 Lawangs. The tide or current being very strong, we reached Lubong Angin and I succeeded with two of my men in reaching the top of the mountain. It is composed of Limestone, contains many caves and the surface is covered with small blocks of Limestone, the edges and surfaces of which are sharp as knives. It is in the highest part about 300 feet. I found three more of the Rhododendron brookeanum, 2 of them epiphytal, on trees; one on a rock with its roots scarcely covered by dead leaves. I obtained about a dozen more of the Anoectocheilus. It was growing amongst moss on the old trunks of trees, very vigorous but in small plant.s I found two or three other pretty plants of a small hairy shell. I obtained some specimens but though the mountain was strewed with dead ones of many species I could not find the retreats of the live ones although I sought diligently for upwards of two hours. I killed a large green hammer headed viper on a bush. The deadly beast was in act to strike had we unwarily approached him, but we saw him at the distance of a yard or two and I cut him in two with my parang. He is said to be if possible more venomous than the cobra and is certainly more feared by the natives. Unless I had killed this one my two men would have allowed him to live. At 4 PM we dined at the place where the antimony ore is conveyed in boats to Sarawak. Soon after passed Seniawan [and] at 11 1/2 PM reached Sarawak.

Nov 28. Employed arranging the plants I brought from the Interior in the glass cases.

Nov 29. Was told by Mr Brooke that he receives very bad accts. of Patingi Abdul Rahman's behaviour at Serikei and that Pangiran Usup and Yacub (brothers) had been killed at Kemanis but the orders from Borneo were to kill them. Usuf requested that he might be bowstrong according to the practice with noble offenders and in this they obliged him. Borneo will now be quiet. I bought a Pheasant and a crested quail today.

Nov 30. Attended prayers at Mr Brooke's and made arrangements for ascending the left hand River tomorrow morning.

Dec 1. Left Sarawak at 2 PM for an excursion up the left hand river. At dark arrived at Karangan Landi above the River Staat. Here we cooked the dinner and passed the night.

Dec 2. Started again at day light and passed in succession the village of Dampul and the Suntah River. At 8 marked the village of Bajack where we stopped to breakfast it [illeg.] passed the Rheum Panjang and Besar, soon after the village at the foot of Gunong Sebayat and Gigi the head quarters of the Sempro tribe. At ten we reached the village of the Sibonoyoh Dyaks where we stayed for the night. I did not ascend to the houses but the Dyaks brought down fruit to the boat.

[Dec] 3. Left Sebonyoh early and breakfasted at the landing place of the Brang Dyaks. We heard that the Dyak tribe of Brang was under the Pamali and that we could not go up till to-morrow and as I was unwell I did not attempt the three hours walk from the landing place but pushed to Penkallum Ampat having first passed the dangerous Rheum. Lodong Ampat is the landing place for several Dyak Tribes of which the Sedong, Goon, Tabiah, Baddat and Sepanjang are the principal. We tried to get up the Senat branch of the River but found it too shallow to admit the boat. We returned in the evening and anchored a little above the Ledong where I collected the Ixora I found on a former trip to Sempro. There were no seeds but I got some young plants.

[Dec] 4. Passed the Ledong without accident and soon reached the Brang landing place when we breakfasted as I intended to visit the tribe, but as we were preparing to go up a letter was brought me from Mr Williamson requested me to assist the bearer in collecting a debt for the Tuan-Ku Sereib Hussin from the Tumma Dyaks. As I had promised to assist the Tuan-Ku if he had Mr Brooke's permission and much rain having fallen during the night I agreed to return and again by the Seriat

River. We did so and with some difficulty reached the village at about 3 PM. All the men were away at their plantations and with the exception of the Orang Kaya Bye Ringate who was under the Pamali Peniakit on acct. of a relation who was very ill, we waited in the boat till some of the old men came home when we went up to the Pangah to dine till the Chiefs of Seniawan could be called. During the evening the men of the tribe came to see me, each bringing in his hand a present of Rice, Fruit, Fowls, eggs etc. The Orang Kaya Pa-Moon was away and By Ringate sent Pa Benang and Pa Pata to enquire if I had any bechara [news]. I told them what I had come for and requested them to appoint a man to call Pa Punang and Pa Mayang Chiefs of Tumma. This they promised to do next day.

Dec. 5. This morning Pa Pata started for Tumma which is half a day's journey from Senah. I employed myself in looking about the village and observing the Fruit trees, the principal of which were the Krakak, Langyir, Kinan, Rhambutan, Barangan, Mangoes, Tampui, Durian, Blimbing, Champadak, Nangka, Langsat, Tancallah, Parrit pisang etc. These were all now ripe and had a beautiful appearance. In one tree I observed a small hut about 30 feet from the ground and on enquiry respecting its tenant was told that it was occupied by a woman afflicted with some loathsome disease which they call rottenness. She had lost many of her fingers and toes and other parts of her body had fallen away. She had been thus afflicted and confined for two years and was fed in her aerial habitation by her relatives. She had born children while still in health. In the evening I made enquiries respecting the customs of these Dyaks. In the first place respecting their dead, the bodies of the dead are burnt not later than two days after decease upon a pile of large logs, logs of wood being also placed around the above body then set on fire by the Baleam (who also collects the materials) in the presence of the relatives who remain on the spot till the whole is consumed. If during the fire the smoke ascends directly towards heaven it is remarked as a good omen, but if from any cause it remains near the earth or flies off in a slanting direction [it] is considered a certain sign that someone present is soon to follow the deceased. The firing having been completed, the friends all return to the house and make a feast. The house for twelve days is under the Pamali and no one but the relatives and occupiers of the house can enter it if the dead man was the father of a family. The custom is that for four days after the burning boiled rice is spread along the house in a line from the door to the opposite window; this the Spirit of the dead is supposed to eat. After 4 days and nights a basin is broken outside the door, the rice in the passage is discontinued and the Spirit who had remained near the House during that time is supposed to go away, to what place they appear to have no idea, and the goods of the person after death are divided first into two portions, that is to say, the household goods such as basins, gongs, tambuk [hearths] and other utensils. The other portion is that belonging to the Spirit of the deceased which is broken to pieces and deposited near the place where he was burned. The other, together [with] everything remaining of the dead man's property such as fruit trees padi either standing or in store even to the posts of the House, and apportioned equally amongst the Children, nor is any distinction of sex recognised. On enquiring respecting their social connections, I learned that it is not considered at all wrong for young women to cohabit with young men. It is generally practised through the tribe, nor do the young women grant their favours to the single men but to as many as they choose. After betrothal or marriage it is different and considered a great crime, indeed one which is never committed so great is their abhorrence of the practice. For the most part the only ceremony in marriage is asking the girl of her parents when, if they have nothing to object and the girl herself consents, it is considered a marriage and they hereafter live together. The parents never thwart the wishes of the young people, nor is it considered presumption in the son of the poorest man demanding the daughter of the richest in the village. They follow the maxim that industry will soon create [illeg.]. In the case of the more honorable men of the village, they are generally betrothed from two to three months before it is called a marriage but during this time the betrothed pair live together as if the marriage where [sic] already completed excepting that it is in the house of the father of the bride. The marriage is considered complete after a pig has been killed and all the village invited to a feast. The bride is then taken to her husband's house and they are looked upon as man and wife. They marry at all ages according to the inclination of the young people whose wishes alone seem to be studied by the elders of the village. The girls are generally very young. There is no difficulty in a man and woman separating. If either or both feel so inclined they have merely to inform the Head Man that it is their intention to do so and each can marry again as soon as he or she feels inclined, but it is not often that they thus separate. I had previously observed that when feats are made a portion of everything at the feast is placed on a neat little stage of Bamboo outside the principal door of the village. Over it is generally a young shoot of Bamboo which with them seems a sacred plant. I have often enquired and I now give the substance of the information I have gained. When a feast is made for a peaceable purpose they invoke the Blessing of their God Juwata (who however from all I can learn they have got from the Malays as they generally call him Juwata Sant) Matahari Bulan Bintang, Rajah Bruni and the Tuan Besar. The portion of food that is placed outside is for Juwata, a God who resides in the Skies, a God of beneficence and peace. He appears with the Sun, Moon and Stars to [illeg.] the seasons as he is always invoked for weather suitable for their Rice crops. He is supposed to extract during the night the substance or essence of the rice, pork etc. and on the following morning it is eaten by the Dyaks. On occasions connected with war and bloodshed they call upon other gods and neither of those above mentioned are at all consulted. Trin and Kamang who reside on the highest peaks of the most lofty mountains are the martial Genii of these people. The former is said in form and features to resemble a Dyak himself [illeg.] than any one can conceive, inspiring his votaries with valour. Kamang is a ferocious and sanguinary God, delighting in feeding on the blood of human beings, particularly the enemies of those tribes who worship him. If the slight reverence they pay to any of these Deities deserve so to be called, Kamang is supposed to be in the highest degree ugly and covered with shaggy hair. These two spirits of whom Trin is the most powerful are consulted before a war is undertaken by the tribe. If they wish the War and design their votaries to be successful, they descend from the mountains and make known their wishes and presence to the Dyaks by kicking the posts of the houses till they shake upon the heads of the Dyaks. This having been heard and the omens of the birds, insects etc. being favorable, they undertake the expedition in the certain prospect of success, supposing their Gods to be with them and having returned successful at the end of the year when the standing crop of padi is gathered a great feast is made. The two largest pigs in the village are killed, large quantities of rice are cooked and [illeg.] of Toddy are prepared. The whole tribe then for from two to four days give themselves up to intemperance and intoxication, women and children not excepted, nor amidst the general festivity are Trin and Kamang forgotten. Their share is suspended as in Juwata's on peaceful occasions but with this difference that whereas Juwata's is eaten after one day and night that of the Gods of War remain for four days and four nights and though when putrid is eaten by the people as a sacred morsel connected with their religion and superstition. The Pamali must be mentioned. As far as I could learn in this tribe, it is of three Kinds (but I think that it is amongst other tribes extended to more and probably with this one also), the Pamali Mati, or the Taboo for the dead, the Pamali Peniakit or that for sickness and the Pamali Omar or that of the Padi plantation. The first Pamali Mati is on a house and on everything in it for 12 days after the decease of any person belonging to it. During this time no strangers can enter the house, nor can the persons belonging to it speak to them, nor can anything be removed from it on acct. until the prohibition has expired. The Pamali Peniakit is undertaken by a relative in favor of a sick person and is a kind of [medicine] practice. A father supposes that by debaring himself from all society for 8 days or less according to the severity of the disease that his son will recover. He may not during the Pamali speak to anyone who does not live in his own company an during the time the House and utensils are tabooed. The Pamali Omar lasts four days. It is on the plantation. As soon as the padi is all planted and during the four days and nights no person for any purpose, not even the owner of the farm, can set his foot in it. The time having expired, they kill a pig and make a feast as on other occasions of the Pamali. There are I believe circumstances under which whole tribes are 'pamali 'so that no one can go to the villages, but I have not yet had opportunities of making good enquiries.

Dec 6. In the evening the Tumma Chiefs with the Chiefs of Semang Kun and several other tribes who have heard of my being at Senah came to visit me. I laid the little business before By Ringate (who had broken through the Pamali) and the Tumma men they agreed to pay the rice and merely requested their ten pasus of it might be left until the next harvest was collected, which I said they might do as I did not suppose the Tuanku would run them hard. They made a feast today to which I descended walking on Gongs [illeg.] to a kind of Couch they had made here they presented me with abundance of Rice, Fowls, Eggs and I went in the evening. The men and afterwards some women danced in their usual ungraceful style. The young women were too modest to be seen and none of my Malays were allowed to enter the Houses whilst the women were there. It has not been so at any other Dyak tribe I have visited.

Dec 7. At 9 AM left Senah. On our return there was a fresh in the river so that we came down with great [speed], accompanied by some danger. On our way I collected many seeds of the fine Ixora but got no other plants on this expedition. I brought with me 3 hives of bees which I had observed amongst the Senah Dyaks. No other Dyaks keep them in hives as far as I am able to learn. They are small [and] of two kinds, brown and yellow, and said to produce much Honey.

Dec 8. At 3 AM we reached home unwell with a slight fever and Dysentery.

Dec 9. Unwell but went to the house. Mulana of Kalekka and the Orang Kaya Permancha of Seribas arrived to-day. Some of the Seribas Chiefs, Dyak and Malay, have been here and gone again during my excursion.

Dec 10. Still unwell but again went out with Mulana and the Orang Kaya in the evening at Mr Brooke's.

Dec 11 and 12. Unwell. Can't get about without much pain.

Dec 13, 14, 15. Still unwell.

Dec 16. Caught by my men a young pig in snares I had set for pheasants. It was striped down the back with a light honey colored [illeg.] covered with reddish brown hair.

Dec 17. Caught in the snares 3 partridges and another pig. Found in Hentig's clearing a new and beautiful Orchidaceous plant; it flowers in a spike from the top of a neat 4 sided 2 leaved pseudo bulb, the flower purple and smells like wormwood. Not yet sufficiently recovered to get about.

Dec 18. The Springes [traps] retained a porcupine this morning. It seems precisely as far as I can judge from recollection to resemble the porcupine of Europe. They had cut its head nearly off and one of its legs so that its skin could be of no use to me. This they did that they might eat it. A [illeg.] was caught and some large animal had escaped

Dec 19, 20, 21. Laid on the shelf altogether. The Dr attending me and Mr Brooke sends me books and calls himself to see me every day.

Dec 22. Much better. My men brought me a plant of a new Pavetta. It was so large that I could do nothing with it except take some cuttings which I have planted in a glass case.

Dec 23. Ill again at home.

Dec 24. Little better. Went to Mr Brooke's to dinner.

Dec 25. Better.

Dec 26, 27, 28, 29, 30, 31. Still keeping better but so weak that I can not walk a hundred yards.

1846

Jany 1, 2, 3, 4, 5. Still very weak and at time[s] feverish. On the 1st we had boat races which were well contested and attended.

Jany 5. Went down to Santubong for [a] change of air and at night was nearly killed by mosquitoes and sandfly.

Jany 6. Cut down a Durian Bruning which is very different from the cultivated kind in the largeness of its leaves and the spherical shape of its fruit. Its stem was 70 feet to the first branch. It was a young specimen. At night again troubled with the mosquitoes.

Jany 8. Cut down tree which produces the Dammar Mata Kuching. The wood was very hard and the [trunk] 70 feet long. In the afternoon felled a Kappur Barus but there was no camphor in it being a young tree wood exceedingly hard stem 60 feet to the branches. Find myself stronger and being in a fever from mosquito bites I determined to go home with the flood to night, but having nothing to eat but biscuit we pulled over to Monkey Island to look for pigeon. I killed five pair (C maritima) on which we dined comfortably uppon Monkey Island. On the Main [land] I found an Ixora which appears to be a variety of that one found in the neighbourhood of Sarawak. It grows with a single straight stem 16 to 18 feet high. The leaves are smaller and the branches more numerous and twiggy. I gathered a few seeds of it for it is pretty.

Jany 9. Arrived at Kuching early this morning. Made a lazy day of it.

Jany 10. Hard at work all day preparing part of my specimens of woods for transmission to England. Find them in pretty good condition.

Jany 11. Sunday. Hentig prayed and dined with us in the evening. Had a long discussion about poetry, a favorite theme with most of us.

Jany 12. Monday. Sent my men out by themselves whilst I wrote part of a very long letter to Clapton detailing my success up the river last month.

Jany 13. In the morning searching through the settlement for old boxes to send away plants by the Schooner PM. Searching for Orchidea and found a small plant of the fine Cypripedium. Saw others on a tree and required my men to fell it and obtain them to-morrow. Examined the seed pods of a Rhododendron Brookei. They are in a fine state but will not be ripe for a month.

Jany 14. Sent Cumian, one of my men, to the Banjar Kalamantan to examine its seed pods. They are not yet ripe. I accompanied the men who went to look for the Cypripedium and it gave me an opportunity of seeing the agility of the Dyaks in climbing. The tree was high and far too large for them to span with their arms. Nevertheless they climbed by the assistance of their fingers and toes. We got 6 good plants of the Cypripedium and found a very pretty but minute orange flowering Orchideous plant on the same tree.

Jany 15. Hoped to have gone up the River to day in search of the [illeg.] which abound above Ledah Tenah on both Rivers but my men disappointed me and could not be induced to go.

Jany 16. The damned rascals won't go. They are all going away on various pretences and though I spent the morning till an hour past noon in looking for men I fear I shall not get away. Every one is a trader on his own acct. and does not wish to work for wages. Curse their proud stomachs. At three PM having made a push in my boat up the village by calling at every Campong, I succeeded in getting a crew though none of them are clever at our river work they being all strangers. Though the tide had flown for some time by great exertion we reached Landi at 8 PM where we took my old station under the trees for the night.

Jany 17. At 8 in the morning reached Suntah. The diamond River has lately overflowed its banks which are 16 or 18 feet above the ordinary limit of the stream. The Nutmegs are all dying. I can't tell the reason of [sic] it; they were in a flourishing condition two months ago, growing with strong shoots like willows. The third crop of Durians, Langsats and other fruits is now ripening. They are in the greatest possible abundance.

In the afternoon I went up the Diamond River in a small boat to look for Orchidaeous plants but was not successful in finding any new ones. I passed a place where I was told the Dyaks buried the bones of their dead after the bodies were burnt. It was an enclosed space where clusters of beautiful Nibungs were growing. I was further informed that where this species of Areca is found the place is sacred and cannot be broken up for farms or other purposes.

Jany 18. Sent a man to the Dyak Houses to try and buy fowls and eggs. Went myself into the Jungle to look for flowers. I followed the course of a small stream as I had seen small plants of a trichomanes on it which appeared different. I had not gone far before I came to some old ditches which appeared to have belonged to diamond workers. I soon after saw one fresh opened and the diamond soil exposed. On still following up the stream the ditches became more numerous and at last I saw the troughs and dams used for washing them I found on enquiry that they belonged to Liby Jamal who had had Mr Brooke's permission to work for three months. He must have had a great many more men than Mr Brooke was aware of to have carried on his works so extensively. When the time had expired he said he had found but few diamonds and had lost by the speculation but was very anxious that Mr Brooke would give it him again. But this could not be done unless he paid a tax but this he has no money to do. In the evening dropped down to the main river that I might catch the ebb at Ledah Tenah by starting early tomorrow. One of my men I left at Suntah to set traps or snares for pheasants, partridges, Quails etc.

Jany 19. Arrived at Kuching at 2 1/2 H PM found that Mr Brooke had purchased me an Argus Pheasant, a fine male, for five rupees. I suppose he will not live long.

Jany 20. Slightly unwell. Skinned one of my pheasants which died last night.

Jany 21. Employed about my specimens. My men have been ill for the last 7 days so that I fear I shall not get any plants ready by the [time the] Schooner comes as my own weakness and illness prevented my doing it earlier. Some of my plants in the glass cases look well but all want replanting before they can be sent away.

Jany 22. A slight cut inflicted on my left hand while skinning the pheasant the other day has swelled and put my whole arm in pain (probably some of the soap got into the wound). The Dr advised hot fomentations, poultices and no wine (I am out of cigars) as he says from such a cause it may endanger my life. I have put myself under his care and spent the day with the fomentations which considerably relieved the pain.

Jany 23rd. Fomentation poultices and medicine continued and the hand still keeping better. The pain has entirely left the arm about midday I was surprised by seeing Bulan, the Sakarran Chief, enter my house. He had just arrived and brought me a present of some Langsats. He told me that he expected the other Sakarran Chiefs with Abang Kapi immediately.

Jany 24. The swelling has abated on my hand in a great measure so that I went to the House to Breakfast. As we were rising, Abang Kapi's boat arrived. They brought a fine male Mias Pappon and a young bird for Mr Brooke. In the evening after dinner there came up to the House Gasing and Lingi. [They] are the fresh arrivals with Kapi [and] of consequence they are attended by other Chiefs and have merely come on a friendly visit.

Jany 25 Sunday. In the morning attended prayers at the House. Afterwards went to see how the Rhododendron seeds were getting on. They are nearly ripe and will yield a fine though small harvest. Returning, I arranged in part two glass cases of plants to be sent by the Schooner. In one is a magnificent plant of the Rhododendron, in the other is the Anoectocheilus, the Stapeliaceous plant and Aeschynanthus Aucklandiae, all established and growing well. I hope to God they will reach home safely. They will have a good chance. My hand and other sickness has prevented me from preparing so many cases for England this time as I wished, but as the Schooner has not yet arrived if my men get well I have yet hopes.

On asking the Sakarran Dyaks about their customs of marriage they told [me] that they could not marry sisters or cousins or even foster children if they have been brought up by the same nurse, though they have no relationship. An uncle cannot marry his niece and the reason they say all this was that which holds with us, that it degenerates the species. I have got only one pen and it is such a damned bad one that I can't write [anything] else. I had more to say.

Jany 26. Either working at the boxes or the wine at dinner last night has made my hand as bad as ever and is worse for the seat of pain has removed. I have been poulticing it all day. Understanding that Mr Brooke is going to instal the Sakarran Chiefs. I don't know if he has any object in doing it more than strengthening a good cause, but it seems that they will consider they are his deputies and not those of the Sultan of Bruni.

Jany 27. My hand is slightly better. My man has come down from Suntah [and] the other two are ready to go to work. No birds from the Diamond district except the green red crested partridges. My Argus Pheasant is doing well. I sent the men to the Jungle.

Jany 30. This morning Crookshank and myself were employed in endeavouring to find the remains of poor Williamson who fell overboard and was drowned last night about 10 PM as he was returning from dinner. We returned, as we expected we should, unsuccessful and from the number of crocodiles we saw I have very little doubt that the body has been entombed in the stomachs of one or other of these voracious animals. This death has visited in an awful guise our small society and taken from it by far the most amiable (Mr Brooke excepted) person of those composing it. From the beginning he had been Mr Brooke's chief assistant and the esteem he gained for himself in the discharge of his many duties is acknowledged no less by the natives than by his superior and his friends. We all loved him and each lament him; in deed, of all here he is probably the only one whom I had considered as more than an acquaintance. His harmless life on earth and his happy temper and disposition leave no doubt that that his sudden death must have introduced him to a yet more blissful state.

March 28. HMS Hazard arrived from China. She has been off the Borneo River and brought here Beder-ed-din's personal slave who conveys to us the melancholy intelligence of his master's murder, together with the Rajah Muda Hassim and nearly all the other of the royal brothers. The crime was perpetrated about 2 months since by the surviving relatives of Pangiran Usuf who had been caught and killed at Kemanis soon after Sir Thos Cochrane drove him from Borneo in the autumn. Beder-ed-din is reported to have died heroically when he found his house surrounded and on fire. He fought with desperate valour but overpowered by numbers, a severe wound across the chest and right shoulder disabled him from using his weapon any longer so that to prevent any barbarian triumph over his remains he threw himself into his burning habitation and perished in the flames. The Rajah too behaved gallantly and was likewise burnt in his house. Thus has perished all that was good in Borneo. The infamous Sultan is now surrounded by ministers more wicked than himself. They are fortifying the heights in fear of the English avenging those who died Martyrs to the interests of Britain. We hear that Sereib Shaif has fortified a position at the mouth of the River and expresses his determination to hold it against the assassins until the

English come. It is a proverb that bad news seldom comes alone. 10 days since we had the Chief of Lingah here. His place at Banting had been attacked from the Batang Lupar by a force of 70 Dyak Bankongs. They burnt the Houses at the foot of the hill and killed and took prisoner 33 people, Balow Dyaks and Malays. Bulan, whose son is on board the Schooner and who was here [illeg.] months since promising fidelity, was amongst the most active of the assailants.

HUGH LOW'S
PLANT PORTRAITS

Phillip Cribb
The Royal Botanic Gardens,
Kew

Established in 1820 at Upper Clapton about 8 miles to the north-east of the City of London, the nursery of Hugh Low & Co was one of the first to successfully import and grow tropical plants in England. In 1843 and 1844, Thomas Lobb (1820–1894) had collected many exotic and horticulturally desirable plants in Java and the adjacent islands for Messrs James Veitch & Sons, Low's main rival in London. In response, Hugh Low sent his eldest son and namesake to the Far East in 1844 to collect exotic plants for the family business. The Malay Archipelago, particularly Borneo, was the target that he set his son from the start.

Hugh Low reached Singapore in November 1844 and began almost immediately to collect exotic plants for the family nursery. His earliest entry in his diary was on November 19th–22nd 1844 when he went ashore on Pulau Chalot (Copper Island) and noted that he 'saw many plants which I had known in England, amongst them a *Hoya*, *Dendrobium crumenatum*, a flaming pink *Dendrobium ?secundum* and the Bird's nest fern are seen in great abundance'. By December 5th he was collecting pitcher plants, *Nepenthes*, on Singapore Island and on the small islands offshore. On the latter they grew 'in abundance and in some parts out of the debris of the rock but in such situations did not appear to thrive so well as when growing in a rich soil and in that moist position the two species which Mr C. bought to England were both there but only that with the bright shaped patches was in flower'. The following day, he packed his first case of pitcher plants to be sent back to his father's nursery. This and a further three cases were despatched home on December 17th on the *Chieftain*. On December 18th he found a new *Nepenthes* whose 'cups are small and green but most delicately and beautifully formed'. His drawing of *Nepenthes gracilis* fits this description well and since both *N. rafflesiana* and *N. ampullaria*, the other two species that he drew are also found there, it is possible that all of his *Nepenthes* drawing were made while he was in Singapore. A further case of *Nepenthes* and also collections of fruit tree seedlings were despatched on the *Georgetown* at the end of the month.

He left Singapore for Borneo on January 6th aboard the schooner *Julia* and had reached the coast of Borneo by the 9th, although bad weather delayed them on their approach to the Sarawak River. They arrived in Kuching on the 14th when they dined with Rajah James Brooke. In Borneo, he could scarcely have chosen a more productive area in which to collect plants. The island is blessed with some of the world's most spectacular plants, gigantic forest trees in immense profusion, fabulous palms such as the scarlet-stemmed *Cyrtostachys renda*, several *Rafflesia* spp. with gigantic flowers, diverse pitcher plants including the magnificent *Nepenthes lowii* and *N. rajah*, both discovered by Low, impressive and sinister giant arum lilies (*Amorphophallus* spp.), and numerous showy orchids.

Hugh Low's diary between January 1845 and March 1846 contains many references to the plants that he saw and collected on his travels in Borneo. However, only once did he mention drawing a plant and that was on January 26th 1845 when he noted: 'Still raining heavily so I amused myself with drawing one of the Ixoras of yesterday'. This may well be the painting of *Ixora pyrantha* (Plate 66) now in the Kew collection which is faintly annotated in pencil 'common in the jungle, flowering all the year'.

One of the most distinguished of his discoveries in these early days was that of a fine epiphytic *Rhododendron* with large trusses of yellow flowers. He first found it on October 24th 1845 'a *Rhododendron* with yellow flowers, epiphytic on the stem of another tree. The bunch is magnificently large and the leaf large and smooth, habit dwarf and drooping'. By November 20th he had named it *'Rhododendron brookeanum* in compliment to James Brooke Esq. of this place whose kindness about him deserves a more lasting monument than a flower can bestow'. Thereafter, he collected several more plants for eventual transportation to England in Wardian cases, in effect little glasshouses that were the more effective means of transporting living tropical plants by sea to Europe. On January 25th he arranged 'two glass cases of plants to be sent by the schooner. In one is a magnificent plant of the *Rhododendron*, in the other is the *Anoectochilus* [a terrestrial jewel orchid]'.

Little is known of his subsequent collecting, apart from the Borneo plants that Hugh Low & Co were able to offer their clients during the 1840s. Low returned to England in 1847 but in 1848 took up a post in the new government of Labuan. From that time onwards, plant hunting became a recreation and his large shipments of exotic plants to his father effectively ceased.

The position that he took up was as Colonial Secretary on Labuan. He was based for many years on this small island off the northwest coast of Borneo and an ideal spot from which he could make forays into the adjacent Borneo coast and hinterland. The most famous of these was his ascent of Mt Kinabalu in 1851, being the first white man to reach the summit. Thomas Lobb, collecting for Messrs James Veitch & Sons of Chelsea, ascended the mountain five years later only to be turned back before he reached the summit by local natives who feared that he would disturb the gods that lived there. Low's interest in plants was well-known to the native peoples and his command of their language and familiarity with their customs gained him widespread admiration and respect. He also took an interest in the agriculture of the region which was particularly noted for its spices and fruit trees. No doubt, he also sought to improve the food supply on Labuan and he founded a garden around his house. Fruit trees and ornamental trees and bushes would have been integral to the garden. It seems likely that the plants in the garden provided the materials for what are probably his drawings of a selection of native and exotic fruit trees, spices and ornamental plants. He certainly collected fruit trees for his family firm in his early days and detailed the fruits that he found. Thus on December 6th 1845 we find him describing the fruit trees around the village of Seniawan that included 'krakak, langyir, kinan, rambutan, barangan, mangoes, tampui, durian, blimbing, champadak, nangka, tancallah, parrit pisang etc.'.

Low's interest in plants was well-known to the native peoples of Borneo and his command of their language and familiarity with their customs gained him widespread admiration and respect. The jewel orchids *Ludisia discolor* and *Anoectochilus* spp. became known as *bunga lo* (Low's plant) or *daun lo* (Low's leaf) by the people in west Borneo and the Anambas and Natuna islands.

HUGH LOW'S ORCHIDS

Even before landing at Singapore, Low had seen his first epiphytic orchids on Copper Island. On December 18th 1844, he went with Mr Little to Mr Armstrong's plantation in Singapore to 'look at some trees which we were informed were covered with orchids. Lost ourselves in a tigerish looking jungle and got out with great difficulty'. On December 31st he found orchids on mangrove trees.

Orchids feature frequently in his Borneo diaries of 1845 and 1846. Indeed the orchid flora of Borneo alone, at over 1500 species, is richer than the entire flora of the British Isles.

LOW'S SPECTACULAR DISCOVERIES

Low's early collections in Borneo caused a minor sensation when they arrived back in England. They included the lime green and black-flowered epiphytic orchid *Coelogyne pandurata* and its close relative *C. asperata*. Amongst his consignments, the most notable were those that were later named in his honour. Pride of place must go to the plant that is now called **Dimorphorchis lowii** (Plate 1), but which H.G. Reichenbach first named in the *Gardeners' Chronicle* in 1847 as *Vanda lowii* based upon Hugh Low's first collection. It is an imposing and, at the same time, peculiar orchid, the only species of orchid in the Old World to have two different kinds of flower in the same inflorescence. Low described it well in his diary of November 23rd 1845. 'Soon I observed an orchideous plant on the top of a high tree and sent a man after it. It proved to be an immense mass of *Vanda* like plant which I have previously sent home though not seen the flower. On this tree were about 100 drooping spikes of flower each from 9 to 10 feet long; they arise from amongst the leaves which clasp the stem at their base, generally one raceme of flowers to each stem but very frequently more. The flower is more than 3 inches in diameter, light sulphur yellow and blotched very much with rich cinnamon brown. The labellum is purple in the centre and spotted with brown towards the edges. The flowers as they get old do not fall off for a very long time but gradually lose their rich blotches and acquire their yellow colour so that the first flower on the stem by the time the others at the end have opened is of the richest golden yellow. The flower stem is rough and the pedicels of the flowers are protected by a sheath above their upper surface as long as themselves. They are thickly covered with a brown substance resembling the moss of the rose. The leaves are of a light green colour and leathery texture, a little more stiff than in *Aerides odorata* of the same shape 18 in. to 2 feet long. The roots do not arise amongst the leaves but are wholly confined to the bottom of the stems where the leaves have long since fallen. They are large but not so thick as might have been expected from the size of the plant, the stems are 2 to 6 feet long and grow horizontally from the tree. The scythes of flowers hang down below.' In all this Low was a careful observer except that the basal two flowers of each inflorescence are quite different from the rest, being yellow spotted with red and highly scented, while the rest are of a larger and are cream coloured with large red blotches all over the sepals and petals. To this day, the reason for this floral dimorphism is unknown but must be related to the pollination.

Almost equally impressive is Low's slipper orchid, **Paphiopedilum lowii** (Plate 2) (originally named as *Cypripedium lowii*), which he first discovered in Sarawak on January 13th 1846. It is one of the few slipper orchids that is epiphytic in habit, growing on tall forest trees, often by streams and rivers. It is an impressive sight when in full flower. An individual plant can comprise several fan-shaped growths, each with several strap-like glossy green leaves. The inflorescences, up to a metre or more long, arch out from these fans and carry between three and seven flowers each. The flowers, up to 15 cm across, have a characteristic slipper-shaped lip and spreading spoon-shaped petals with purple tips and dark blackish spots in their lower parts. Low described its discovery thus:

'searching for *Orchidaea*…found a small plant of a fine *Cypripedium*. Saw others on a tree and required my men to fell it and obtain them tomorrow'. The following day they climbed the tree which was 'high and far too large for them to span with their arms. Nevertheless, they climbed by the assistance of their fingers and toes. We got 6 good plants of the *Cypripedium*'. We know that Low painted his fine discovery but the painting is sadly lost. John Lindley, the assistant secretary of the Horticultural Society of London and the leading orchid taxonomist of the day, described it in 1847 as *Cypripedium lowii*. His original collection is at Kew in John Lindley's herbarium and John Day's painting (also at Kew) of one of the early introductions of this fine orchid is reproduced here. (Plate 2).

In *Dendrobium*, a genus with many spectacular species, Low discovered one of the finest. **Dendrobium lowii** (Plate 3) is in a group of species with stems that are covered in short black hairs. Most of its allies have white flowers variously marked in the lip but Low's discovery has large bright buttercup yellow flowers with a beard of red hairs on the lip. A long slender spur-like mentum extrudes from the back of each flower. It was described in the *Gardeners' Chronicle* in 1861 by John Lindley based upon Low's collection from Sarawak. The original type collection is in Lindley's Herbarium at Kew and one of John Day's paintings of it is reproduced here. Even today, it is highly sought after and only a few collections have managed to flower it in recent years.

The fourth of the orchids named after Low is **Plocoglottis lowii** (Plate 4), a strange plant indeed. It is a terrestrial orchid found on the dark forest floor. Its has a fleshy succulent stem bearing a few dark purple leaves. The inflorescence is terminal and the flowers are strangely twisted. In fact, the lip resembles somewhat a fly or bee and on being touched by a pollinator snaps shut against the column, thereby pushing the insect against the pollen masses which adhere to its back. The lip resumes its original position some time after being triggered to catch the next potential pollinator if the first one failed to take the pollen. Low's painting of *Plocoglottis lowii* is labelled as *Plocoglottis* sp., presumably having been painted before the plant was named by H.G. Reichenbach in the *Gardeners' Chronicle* of 1865. It is likely that the painting, now in the Kew archive, is of the type collection.

The only other Bornean orchid named after Low is *Malaxis lowii*, a small terrestrial orchid of little horticultural merit. Eduard Morren described it in 1884 in a Belgian horticultural journal.

LOW'S ORCHID PAINTINGS

Hugh Low painted some 33 orchid portraits while in Borneo. We cannot be sure when he started or finished the series of paintings, none being dated. However, the subjects include a few that are considered endemic to Borneo and it seems likely that he drew them when he was living on Labuan, possibly when he was frustrated at not being able to get across to the mainland to collect. This may also explain why he did not paint many of his finer discoveries such as *Dimorphorchis lowii* or *Dendrobium lowii* which would probably have been unavailable to him on Labuan. His illustration of the pigeon orchid, **Dendrobium crumenatum** (Plate 5 left hand side), also suggests that he drew what was readily available. The pigeon orchid is probably the commonest of all South East Asian orchids and can still be commonly found on the street trees in Singapore and other Malayan and Indonesian cities and towns.

Only two of his paintings are of ground orchids. *Plocoglottis lowii* is a large format portrait (*c*. A3), drawn life size, which necessitated him to paint its flowers on a separate piece of paper and glue it next to the tip of the rhachis on the main sheet. The other terrestrial orchid that he drew was **Calanthe pulchra** (Plate 7), one of the finest of the terrestrial calanthes. Its flowers are a bright orange with a darker orange to almost red lip and with a pretty sigmoid slender spur at the back of the flower. It is widespread in western Malay archipelago and the Malay peninsula as far north as peninsular Thailand. It is found in lower montane, hill and heath forest, often near streams between 600 and 1700 m elevation.

The remaining paintings are of epiphytic species, that is orchids that grow on trees or sometimes on rocks if the conditions are moist enough. Included amongst these are paintings of several genera, including *Appendicula, Bulbophyllum, Dendrobium, Eria, Flickingeria, Luisia, Micropera, Paphiopedilum, Phalaenopsis, Thrixspermum, Trichotosia* and *Trichoglottis*. Nearly all of these drawings were unnamed or named only to genus when they were presented to Kew by Hugh Low. The names now provided are tentative because of the lack of corresponding herbarium specimens whose flowers could be dissected to confirm the identifications.

The majority of the drawings depict species of the three largest genera in Borneo, respectively *Bulbophyllum*, *Dendrobium* and *Eria*. Because of their diversity the species can be difficult to identify and name and Low's drawings lack any detailed dissections. Nevertheless, we can identify three of his four *Bulbophyllum* drawings with some degree of confidence. All belong to section *Cirrhopetalum* which some authorities consider to merit recognition as a distinct genus. The section is characterised by the one-leafed swollen stems (pseudobulbs) and inflorescences in which the flowers are borne in a false umbel, rather like the flowers in a daisy head. Furthermore, the lateral sepals are much longer than the dorsal sepal and petals and are joined for much of their length. Low's drawings depict **Bulbophyllum brienianum** (Rolfe) J.J. Sm. (Plate 8), **B. lepidum** (Blume) J.J. Sm. (Plate 9), **B. purpurascens** Teijsm. & Binn. (Plate 10) and **B. vaginatum** (Lindl.) Rchb.f. (Plate 11). Of these only the last had been named when Low drew it.

Low's eight plates of dendrobiums depict six species. These are respectively two plates of **Dendrobium lamellatum** (Bl.) Lindl. (Plates 12 & 13), two of **D. pachyanthum** Schltr. (Plates 14 & 15) and one each of **D. crumenatum**, **D. tenue** (Plate 6), **D. rosellum** Ridl. (Plate 16) and **D. spurium** (Bl.) J.J. Sm. (Plate 17). The drawing of *D. lamellatum* (Plate 12) is accompanied by a **Thelasis** sp., and one of those of *D. pachyanthum* (Plate 15) by an *Eria* species (possibly **E. neglecta** Ridl.). The most interesting and best-known of these dendrobiums is undoubtedly *D. crumenatum* (the pigeon orchid), so named because the flowers resemble small white doves on a branch. As mentioned above it is a very common and widespread orchid, often found on street trees. It has the unfortunate habit of flowering for only a day and then all the flowers collapse. The eminent botanist Professor Eric Holttum, erstwhile Director of the Singapore Botanic Garden, established that it flowers a few days after heavy rainfall, the drop in temperature that accompanies the rain triggering flowering.

The ephemeral flowering **Flickingeria comata** (Bl.) A.D. Hawkes (Plate 18) is a close relative of the dendrobiums and has been included in that large genus by some authorities in the past. It is a vigorous orchid, often forming large clumps on trees, but its flowers are seldom seen because flowering is synchronous in an area and the flowers last less than a day before withering.

Erias are amongst the least known of all Southeast Asian orchids. Few are cultivated which is a pity because the genus contains an amazing diversity of habit and flower form. The flowers can be produced in abundance but, it has to be admitted that they are mostly small and few are brightly coloured. Nine plates of *Eria* were drawn by Low and they probably represent seven species. The finest of these is without doubt ***Eria ignea*** Rchb.f. (Plates 19, 20), distinguished by its flaming orange bracts and flowers. Low drew it twice under the name *E. cinnabarina* Rolfe (a later synonym). He also drew twice the widespread ***E. pulchella*** Lindl. (Plates 21, 22), an orchid that is found on trees on the beach as well as inland. It may even have grown naturally on trees in his garden in Labuan. Three other plates can be identified with some degree of certainty are of ***E. cepifolia*** Ridl. (Plate 23), ***E. hyacinthoides*** (Bl.) Lindl. (Plate 24) and ***E. linearifolia*** Ames (Plate 25).

The hairy epiphytic genus *Trichotosia* was also once considered to belong in *Eria*, but modern authorities separate it in a distinct genus distinguished by its unswollen leafy and hairy stems and hairy flowers. In most species the hairs are reddish or russet in hue. ***Trichotosia vestita*** (Lindl.) Kraenzl. (Plate 26), the species depicted by Low, hangs from forest trees often in riverine or swamp forest and is found in Borneo, Sumatra and peninsular Malaysia.

Low depicted few orchids of horticultural merit apart from the eponymous *Paphiopedilum lowii*. Exceptions are the moth orchids of which he depicted two species, *Phalaenopsis amabilis* (L.) Bl., the well-known and widely grown white-flowered moth orchid and the somewhat less showy spotted-flowered species ***P. cornu-cervi*** (Breda) Blume & Rchb.f. (Plate 27). The former was, in Low's time, widespread throughout the Malay archipelago, usually growing as an epiphyte on trees overhanging streams or in gulleys in lower montane and hill forest. However, throughout its range it has been collected almost to extinction for horticulture. In Borneo, even local villagers grow it on their balconies or gardens, tying it to trees in around their houses. *Phalaenopsis cornu-cervi*, another widespread species found from India and Burma across to the western islands of the Malay Archipelago, has also suffered a similar fate and is popular despite its less showy flowers because its inflorescences continue to produce flowers one at a time for extended periods.

Hugh Low drew nine other close allies of the *Phalaenopsis*. The first, **Luisia antennifera** Blume (Plate 28), is a strange plant with cylindrical pencil-like leaves and clusters of green and dark violet flowers on the stems. It is one of four species in the genus found in Borneo and is commonly found on trees in plantations near the coast. The second **Micropera fuscolutea** (Lindl.) Garay (Plate 29) is more colourful having larger brownish-yellow flowers in a longer spike. Its range extends from peninsular Malaysia and Singapore to Borneo where it has been found on citrus trees and in hill and riverine forests. He drew three species of *Trichoglottis*, a genus well-represented in Borneo. The first **T. geminata** J.J. Sm. (Plate 30) grows in Borneo on scrub and andesite rocks near the coast, but is widespread elsewhere from the Philippines to Sumatra. **Trichoglottis philippinensis** Lindl. (Plate 5), as its name suggests was discovered in the Philippines, but it also occurs in Borneo and Sumatra. He also drew another species on a composite plate (Plate 31) with a *Thrixspermum*, neither being yet identified to species. The seventh species of the *Phalaenopsis* group illustrated on a plate with *Dendrobium tenue* by Low is **Cleisostoma teretifolium** (Plate 6), a species with characteristic pencil-like leaves and many small flowers in its unbranched inflorescences. It is common in lowland swamp forests.

The final three species in this group are belong to *Thrixspermum*, a genus of orchids with ephemeral flowers. One of these, possibly **T. tinekeae** Schmit. (Plate 31), is on the plate with an unnamed **Trichoglottis**. The other two plates may well represent the widespread **Thrixspermum centipeda** Lour. (Plates 32, 33), a scrambling epiphytic orchid with spider-like flowers. It is common throughout Southeast Asia, the Malay archipelago and the Philippines, growing in a variety of lowland and hill forests, even in the peat-forests of Borneo.

The final plate drawn by Hugh Low is of a rather nondescript and, as yet, unidentified **Appendicula** species (Plate 34), near *A. uncata* Ridl., with small white flowers.

It is puzzling that, with the wealth of showy orchids available to Hugh Low in Borneo, he should have chosen to draw many rather insignificant orchids of little horticultural merit. However, beauty is in the eye of the beholder and I know a number orchid growers and scientists who prefer the small and delicate species to the large and showy ones. Possibly, Low was reacting against the pressures of his family business in choosing to draw these species, or perhaps he wanted to record orchids that he could not identify and name while in Borneo, hoping that the paintings would be sufficient to name them later.

LOW'S OTHER PLANT PORTRAITS

The other plant drawings in the Low collection include examples of fruits, spices, beverages and ornamental plants, reflecting no doubt the interest of his family's nursery in exotic plants. However, the drawings of useful plants might just as well reflect his own interest in growing fruits trees in his garden in Labuan, and in encouraging their cultivation and those of spices, coffee and cocoa in Sarawak. All of his drawings of useful plants are in landscape format on approximately A2 size paper. At Kew there are two sets of similar paintings of many of the fruits and spices. Recently, an American visitor from Moscow, Idaho, brought a third set to Kew which was bound in a landscape format scrapbook and belonged to Mrs Henry Wolcott Balestier who was a descendant of Joseph B. Balestier, the first US consul in Singapore from 1833–1852 (Richard Naskali, pers. comm.). Their similarity to Low's set of drawings and the another set at Kew suggests that sets of drawings of useful and ornamental plants may have been commercially available in Singapore, the artists being local people or possibly Chinese. Certainly, in Hong Kong local Chinese artists drew plants for visiting botanists and collectors and in India, local artists were employed to draw plants for the botanists of the East India Company (Desmond 1992). We suspect that the artists in Singapore were Malay because all of the fruits and spices in the bound set are identified by their Malay names. All of these illustrations were made on landscape more or less folio-sized paper. The smaller drawings in the Kew collection of orchids and ornamental plants are in portrait format and may well have been drawn by Low or for him by a local artist. We know from his diary that he drew an *Ixora* in the forest and the style of that drawing matches that of his remaining small drawings well.

FRUITS

Annona muricata L. (Plate 35) – Soursop (English), sirsak, nangka belanda (Indonesian), durian belanda, durian benggala, durian makkah (Malay)

The soursop is native to the tropical Americas and was introduced into the Philippines by the Spanish not long after the discovery of the Americas by Christopher Columbus. From there it spread rapidly throughout South East Asia and the adjacent islands. It is a small tree, rarely reaching 10 m in height, bearing small greenish yellow flowers and large green 10–20 cm long fruits covered with soft spines and with numerous shiny blackish brown seeds covered in a juicy sweet or somewhat sour white pulp.

The pulp of the fruit is usually eaten fresh as a desert when the fruit is fully ripe, or mixed with ice cream or milk to make a delicious and refreshing drink. However, it is most frequently eaten as a puree after the pulp has been squashed through a sieve. The pulp can be turned into a jelly, juice, nectar or syrup or added as flavouring to ice cream and, in Indonesia, for sweetcakes. In the Philippines, young fruits are used as a vegetable.

In Borneo it can be grown from sea level up to 1000 m elevation and does best in a monsoon climate, flowering in September and October for harvesting in February and March.

Annona squamosa L. (Plate 36) – Sugar-apple (English), sarikaja (Sumatra, Sundanese), sirkaja (Javanese), nona sri kaya, buah nona, sri kaya (Malay)

The sugar-apple is a native of tropical South America where it is widespread. It is commercially cultivated in South East Asia. It is a small tree up to 6 m tall and has a characteristic spherical to heart-shaped fruit up to 15 cm in diameter with a green segmented surface and dark brown seeds surrounded by a sweet creamy white pulp. It is closely related to the cherimoya (*Annona cherimola*) from tropical highlands of Ecuador and Peru but that is not commercially grown in the region. Hybrids between the two are often cultivated but the commercial production of these is in Australia, Hawaii, Florida and Israel.

The pulp from ripe sugar-apple fruits is normally eaten fresh and is also used to flavour ice cream. The green fruits have effective properties as an insecticide and vermicide.

In Borneo it can be grown up to 1000 m elevation. Fruit set is limited to the onset of the rains.

Garcinia mangostana L. (Plate 37) – Mangosteen (English), manggis (Indonesian and Malay)

The mangosteen is probably the most highly prized of all native South East Asian fruits. Its exact origin is uncertain because it is known only as a cultivated species, although allegedly wild plants have been found in peninsular Malaysia but they may be established escapes from cultivation. Mangosteen trees can reach 25 m in height and have a straight trunk with a pyramidal crown. When damaged all parts exude a yellow latex. The flowers are solitary or paired and have four yellow-green petals edged with red. The fruit is hard cased, deep purple and glossy and contains several segments, each juicy white and delicious sweet. The segments taste somewhat of sherbet and each may or more usually does not contain a seed. Functional male flowers have never been found on a mangosteen.

Mangosteen fruits are eaten fresh when ripe, the tough outer casing being split at the end to reveal the white segments. The rind can be used to tan leather and to produce a black dye. Both the rind and the bark have medicinal qualities. The wood of mangosteen is dark red, heavy, coarsely grained and long-lasting. It is used in carpentry and to make rice pounders.

Nephelium lappaceum L. (Plate 38) – Rambutan (English, Indonesian)

The rambutan is a tropical relative of the lychee (*Litchi chinensis*) and its fruits resemble a lychee but have a thicker coat, usually yellow or red, with long soft prickles all over. The Indonesian word 'rambutan' means hairy. The fruit contains a black or dark brown elongated seed covered in an opaque watery sweet flesh. Rambutan trees reach 25 m tall and their wood is hard and dark red, making it suitable for many types of construction.

It is an important fruit throughout South East Asia and rambutans are grown throughout the region, providing there is no marked dry season. In some areas trees will produce two crops a year. It is native to the region and is widely cultivated in the lowlands. About 22 allied wild species of rambutan are found in the region, eight being endemic to the island of Borneo.

Syzygium aqueum (Burm.f.) Alston (as *Eugenia*) (Plate 39). Water apple, bell fruit (English); jambu air, jambu air mawar (Indonesian, Malay).

Water apple is native to South East Asia where it occurs over a wide area below 1200 m. Trees are cultivated in home gardens, often being planted along paths and drives. The tree ranges from 3–10 m tall and often branches low down. The flowers, a mass of yellow-white stamens, are produced early or late in the dry season. Fruits mature 30–40 days after fertilisation. The fruits are red, purple-red, or yellowish white with a glistening waxy skin, bell-shaped and 5–8 cm long when ripe, each contains a single seed. They contain measurable amounts of vitamins A, B1 and B2, and C.

Water apple is a watery, thirst-quenching, sweet fruit that is usually eaten fresh when ripe. In Indonesia it is preserved by pickling in syrup. Various parts of the tree are also used in traditional medicine, possessing an antibiotic activity.

Musa paradisiaca L. (syn. *M. sapientium*) (Plate 40) – Banana, plantain (English), pisang (Indonesian, Malay)

The edible banana is widely cultivated around the world in tropical, subtropical and mediterranean climates. It is generally accepted that cultivated bananas which are mainly triploid (having three sets of chromosomes) arose from wild diploid banana *Musa acuminata* through the development over time of sterility and parthenocarpy (a form of asexual reproduction). It is considered probable that *Musa balbisiana*, another wild banana found throughout the regions has also been involved in the production of modern bananas. Triploids often arise though chromosome doubling of natural hybrids and subsequent crossing of tetraploid to diploid bananas. The plants of the triploids are more vigorous and their fruits lack seeds making them more desirable to eat. The triploid is called *Musa ×paradisiaca*.

Bananas are large herbs and can grow up to 9 m tall, the leaf bases forming a pseudostem topped by a bunch of large oblong leaves up to 4 m long. The fruits are borne on a pendulous inflorescence in bunches of 12–20.

Bananas are either eaten ripe as fresh fruit or they can be cooked and eaten as a vegetable. Edible bananas vary greatly in size and colour. Fruits range from 6–35 cm long and can be yellow, red or green-skinned. Bananas suitable for cooking are generally called plantains in English. They are usually larger than the edible forms, green-skinned and longer. The flesh of a banana can be dried, chipped, pureed or made into an alcoholic beverage or vinegar. The leaves are also widely used to polish floors, line pots, wrap food such as sticky rice, and can be used as fodder and a source of water. The fibres extracted from the pseudostem can be made into cloth.

Bananas are widely grown in Borneo today, either in plantations or around villages or in gardens. They have an ornamental value in gardens as well as providing food.

Ananas comosus (L.) Merr. (Plate 41) – Pineapple (English), nanas (Javanese and Malay), nanas pager (Malay), danas (Sundanese)

The pineapple is another fruit of tropical American origin where it was domesticated well before Columbus set foot in the Americas. It was bought to the Philippines and peninsular Malaysia by the Spanish in the 16th century and has been extensively grown throughout the region since then. Malaysia has a commercial pineapple industry.

The pineapple is an unusual fruit. It is a herb up to 150 cm tall with sword-shaped leaves up to a metre or more long that are produced in a terminal rosette. The flowers are borne in a compact inflorescence, and are arranged in spirals. The fruit develops as the flowers mature by a thickening of the flower stalk and fusion of the individual berry-like fruits. Fruits are more or less cylindrical, usually slightly inflated in the middle and can be up to 25 cm long and weigh 1–2.5 kg. Pineapples reproduce vegetatively. New leaves develop as a crown at the apex of the inflorescence and when cut off can be planted to grow into a new plant. Pineapples also sucker at the base and the suckers can be cut off and planted.

Pineapples are eaten fresh when ripe, the yellow flesh being sweet and very juicy. They are also canned in syrup, the fruits being sliced or cut into segments or pieces. A refreshing juice can be made by crushing the fruits. The fruits can also be crystalised or glacéed. The by-products of canning are used as cattle feed. Other by-products include citric acid, malic acid and ascorbic acid. Pineapple proteases are used to tenderise meat. In the Philippines the fibres from the leaves are used to weave cloth. In Borneo, the Kayan occasionally use the young leaves as a vegetable, cooking them and serving them up with meat dishes.

Artocarpus heterophyllus Lamark (Plate 42) – Jackfruit, jack (English): nangka (Indonesian, Malay).

Jackfruit is one of several edible species of the genus *Artocarpus* (Moraceae) which includes breadfruit and chempadak. It is thought to have originated in the Western Ghats of India but has been widely cultivated throughout tropical Asia for centuries and has become naturalised in many areas. It thrives in warm and wet climates below 1000 m elevation and is widely grown in village gardens throughout South East Asia.

It is an evergreen, medium-sized tree usually up to about 20 m tall with a dense crown and thin leathery obovate or elliptic leaves up to 25 cm long. The inflorescences are solitary and borne on specialised short axillary leafy shoots. The male flowers are borne in barrel-shaped heads, 3–8 cm long, of fertile and sterile flowers. The female flower heads are distal to the male ones and up to 15 cm long. The barrel- or pear-shaped fruit is covered with pyramidal protuberances and grows up to a metre long and 50 cm in diameter. The seeds within are black and covered with a sweet yellowish flesh.

The edible pulp comprises about 50% of the weight of the fruit. When young it is eaten cooked in curries or is pickled in brine and used as a vegetable. Or to make delicacies ('dodol' or 'kulak' in Java), chutney, jam, jelly or paste, or made into candies. The seeds are dried and eaten as nuts or ground into flour for baking. A yellow dye extracted from the wood is used to dye the robes of Buddhist monks. The timber can be used to make furniture and musical instruments. The latex is occasionally used as birdlime.

Artocarpus altilis (Parkinson) Fosberg (Plate 64) – Breadfuit (English); sukun, kelur, timbul (Indonesian); sukur, kelor (Malay).

The breadfruit probably originates in the eastern Malay archipelago and South West Pacific islands but it is now widely distributed throughout the humid tropics. It sprang to fame when Captain Bligh took HMS *Bounty* to the Pacific to collect breadfruit and transport them to the West Indies where they were to be used to feed the slaves on the plantations.

Breadfruit trees grow up to 30 m tall. Their leaves are entire when young but deeply pinnately lobed in mature specimens. The many-flowered inflorescences are axillary, the male ones a drooping club-shaped bearing small yellow flowers, the female ones stiffly upright and globose or cylindrical. The large almost spherical or cylindrical fruits can be eaten immature when sliced and cooked in coconut milk or dried in the sun. The fruits and seeds of ripe fruits can also be eaten. The stored fruit ferments and is converted into a smelly cheese-like paste or ripe. Commercially, the boiled cut fruits are stored in brine. The leaves and fallen fruits make good animal feed. The smooth bark is made into tapa cloth in the Pacific Islands. The light nicely grained wood is used for canoes, toys, crates and boxes.

Tamarindus indica L. (Plate 43) – Tamarind (English); asam, asam jawa, tambaring (Indonesia); asam jawa (Malay).

A member of the legume or bean family, *Tamarindus indica* is probably native to the dry savannahs of tropical Africa but it is widely cultivated throughout tropical Asia, usually in village gardens in Indonesia and Malaysia. It prefers lowland and hill country in the tropics where rainfall is evenly distributed throughout the year or where the dry season is prolonged and pronounced.

It grows to 30 m tall and has a dense widely spreading crown with pinnate leaves bearing up to 16 pairs of leaflets. The showy flowers, borne in lateral and terminal racemes, have cream-coloured lateral and posterior petals with red-brown veins, the two anterior petals being much reduced and white. The fruit is a subcylindrical woody brown pod up to 14 cm long with a thick syrupy blackish brown mesocarp and containing flattened hard brown seeds.

It is a useful tree in several ways. The unripe fruits and flowers are used for souring fish and soup dishes. Trees may produce either sweet or sour fruits. The ripe sweet fruit is usually eaten fresh, the sour ones are made into juice, jam, syrup or candy. Tamarind seeds are also edible after soaking in water and boiling to remove the seed coat. The flour is made into cakes or bread. Seeds can also be roasted and an oil extracted from them which is used in varnishes and paint.

It has many medicinal uses. Amongst these, the bark is astringent and a tonic or used in poultices and lotions to relieve sores, ulcers and rashes. Its ash may be administered internally as a digestive. A sweetened decoction of the leaves can also relieve coughs and fever.

BEVERAGES

Theobroma cacao L. (Plate 44) – Cocoa (English)

Cocoa is the source of both drinking cocoa and of chocolate. It originates in the New World tropics but has been widely grown throughout the tropics since the 19th century. It forms a small tree that begins to crop after about four years but can live to 80 years or more. The tree is unusual in that the small fragrant pinkish flowers are borne in clusters on the stem and branches where the fruits develop after pollination. The woody fruit pods are yellow or red when ripe and when opened show the dull red beans covered in a white mucilage. The beans are scooped out, fermented and dried. Cocoa beans contain about 50% cocoa butter (a fat) and this is largely removed in the process of making cocoa powder for drinking. However, it is supplemented with more cocoa butter, sugar and milk when chocolate is made. Cocoa is a mild stimulant in that it contains caffeine.

Coffea arabica L. (Plate 45) – Coffee (English)

Arabian coffee is one of two species that are widely grown for the production of coffee, the other being *Coffea canephora* (Robusta coffee). Arabian coffee originated in the highlands of Ethiopia and the Yemen but is now widely grown throughout the tropics, the main production areas nowadays being in countries such as Colombia, Costa Rica and Brazil. It grows into a small evergreen tree that produces bunches of white highly fragrant flowers along its branches. The berries are small and reddish when ripe. Arabica coffee produces the best beans in the world, mild, of high quality and greatly esteemed. The bushes are kept pruned so that pickers can harvest the berries without resorting to ladders. The berries are dried, husked and then the beans (seeds) roasted and ground.

SPICES

Piper nigrum L. (Plate 46) – Pepper (English); lada, merika (Indonesian); lada (Malay).

Pepper has been an important commodity since ancient times, being traded across the Old World from India and South East Asia to Europe and China. It is native of the Western Ghats in India but reached South East Asia by 100 BC, brought by Hindu colonists who migrated from India to Indonesia. In the region of Kuching in Sarawak it is cultivated commercially by small-holders who grow the vines up hardwood supports about 3.6 m tall. Alternatively, trees can be used as supports. Pepper grows well on a range of soils and can tolerate a dryish period of up to 3 months. It is generally grown in the lowlands but can thrive up to 1500 m near the equator.

Pepper is a perennial climber that can reach 10 m in height. The inflorescences are spikes bearing up to 150 small bisexual or unisexual flowers which develop into. Flowering in the region occurs in July and extends over a three-month period. The fruits, globose with a pulpy covering that is red at maturity, develop 8–9 months after fertilisation. The seeds are globose and 3–4 mm in diameter.

The fruits are harvested when mature but still green and then air dried to produce black pepper. White pepper is produced by slightly crushing the fruiting spikes, putting them in a gunny bag and soaking them for several days in slow-running water. The peppercorns are then trampled, separated from their spike and covering and dried in the sun for 3–4 days.

Myristica fragrans Houtt. (Plate 47) – Nutmeg (English); pala, pala Banda (Indonesian); pala (Malay).

Nutmeg is one of the most important of all spices, having been traded from the region to Europe as early as the 6th century AD. The exploration and colonisation of India and South East Asia by the Portuguese, Dutch, French and British was driven by the search for sources of cheap nutmeg and other Oriental spices.

The nutmeg tree is the source of both nutmeg and mace, the former the seed and the latter its dried arillate covering. Both are encased in a green fruit that splits in half to reveal the red mace covering the seed. The cone-shaped tree reaches 5–13 m tall and is deciduous. It produces 1–3 flowers in axillary inflorescences at the tops of young branches. The fruits can be harvested throughout the year and a good tree will produce 5000 per annum.

Nutmeg is known only in cultivation but probably originates from the Moluccas. It is still grown in plantations on Banda and neighbouring islands, elsewhere it is mostly grown by small-holders. It prefers a warm climate where the rainfall is steady throughout the year and the temperatures range from 25–30°C. About 20,000 tons are produced annually, mostly in the region.

Nutmeg is used in cooking as a spice or flavouring. Mace is used to flavour savoury dishes. Nutmeg also produces essential oils from the seed, mace, bark and leaves. Extracts are often used in the soft drinks and cosmetics industries. It also has insecticidal, antibiotic and fungicidal properties. Nutmeg also has a narcotic property with hallucinogenic effects and the consumption of two ground nutmegs can cause death.

Syzygium aromaticum (L.) Merrill & Perry (Plate 65) – Clove (English); cengkeh (Indonesian); chengkeh, chingkeh (Malay).

The clove tree was first domesticated in the Moluccas, where it occurs naturally. It also occurs wild in New Guinea. The Chinese have commercially exploited it since at least the third century BC. The cultivation of cloves spread to many of the islands of the region under the Portuguese but when the Dutch ousted them from the Malay archipelago, they destroyed the clove tree plantations on all the islands except Ambon in the Moluccas. However, in the 19th century the cultivation of cloves spread again to Sumatra, Malaya and Sri Lanka thus breaking the Dutch monopoly. In the last century it has spread throughout the region.

The clove tree is a small tree that reaches 20 m tall, conical when young and cylindrical as it matures. The leaves are simple and opposite. The inflorescences are terminal and branched bearing many small flowers usually in groups of three. The flower buds are 1–2 cm long and constitute the cloves when dried before opening. The flowers are yellowish green with a red flush and the fruits an ellipsoidal dark red berry. Clove oil is extracted from the buds.

Cloves have many uses. Their use as a spice for flavouring food drove the Portuguese and Dutch to conquer the Malay archipelago in order to monopolise the trade with Europe. However, more than 90% of clove production nowadays is used to produce 'kretek' cigarettes in Indonesia. Cloves are also important medicinally, e.g., clove oil suppresses toothache and halitosis, and as a stimulant and carminative.

ORNAMENTAL PLANTS

Lepeostegeres beccarii (King) Gamble (Plate 48)

This semi-parasitic shrub is a member of the mistletoe family. It grows on forest trees in the hill and montane forests of Borneo. It was described by Sir George King of the Calcutta Botanical Garden based on a collection made in Sarawak by the celebrated Italian explorer and botanist Odoardo Beccari (1843–1920). The bunch of red-tipped flowers emerging from a cluster of bright red bracts is most distinctive.

Lepidaria sabaensis (Stapf) Tiegh. (Plate 49)

Another member of the mistletoe family, *Lepidaria sabaensis* grows as a semi-parasitic shrub on trees in montane forest and scrubby forest. Low

129

may well have encountered it when he climbed Mt Kinabalu where it is a common parasite on *Leptospermum* bushes and trees from 1400 to 3700 m elevation. It was originally described as *Loranthus sabaensis* in 1894 by the Kew botanist Otto Stapf based on a collection made on Mt Kinabalu by Haviland.

Nepenthes ampullaria Jack (Plate 50)

It is perhaps surprising that Low chose to draw three of the common species of pitcher plants rather than the spectacular ones that he collected on Mt Kinabalu during his 1851 ascent of the mountain. There he discovered the elegant flask-shaped pitchers of *N. lowii*, the large cylindrical ones of *N. edwardsiana*, the hairy ones of *N. villosa* and the enormous ones of *N. rajah*. The last he estimated would carry four pints of water and in one he found a dead rat! However, none of these would have been immediately available to him in Labuan. Instead he drew there three of the common and widespread coastal and lowland species. The depiction of his series of fine montane species was thus left to Walter Hood Fitch who drew them for Sir William Hooker at Kew. It is, however, possible that he drew all of these *Nepenthes* while in Singapore in December 1844 as all grow on the island.

Nepenthes ampullaria is a pretty small-pitchered species and one of the most readily recognised species in the genus. The ground pitchers, usually borne in large clumps on the ground, are very distinctive being green or red and the size of an egg-cup, almost elliptic with a smooth incurved margin to the pouch and a small, almost vestigial recurved lip and are. The aerial pitchers are rather seldom seen but are more obconical in shape. It is a widespread species in Borneo but its range extends from Sumatra, peninsular Thailand and Malaysia to New Guinea. It usually grows in very wet places in peat-swamp forest and kerangas forest in relatively flat areas up to 1000 m elevation. Low would undoubtedly have been very familiar with it in Labuan and on the adjacent mainland of Borneo.

The Kew collection also possesses a watercolour of a ground pitcher of *Nepenthes ampullaria* by Sir James Brooke (Plate 51).

Nepenthes gracilis Korthals (Plates 52, 53, 54)

Another common lowland species in Borneo, *N. gracilis*, as its name suggests, has slender green or less frequently red pitchers. The ground ones are shorter than the elegant aerial ones. It is the commonest lowland species in Singapore and Borneo and is also found in peninsular Malaysia and the western Malay archipelago.

One drawing (Plate 55) may be of the pitcher plant *Nepenthes* ×*trichocarpa*, a natural hybrid of *N. ampullaria* and *N. gracilis* that occasionally can be found in the region.

Nepenthes rafflesiana Jack (Plates 56, 57, 58)

Nepenthes rafflesiana commemorates Sir Stamford Raffles, the founder of Singapore and a fine naturalist who collected plants and animals in Java, Sumatra and Singapore. It is an elegant species. It was described in 1835 based upon a Raffles collection.

It is a common and widespread species from Peninsular Malaysia to Borneo and Sumatra. It grows in wet open areas, kerangas forest and in peat-swamp forest from sea level up to about 1200 m elevation. Its ground pitchers are pot-shaped and bulbous and have two bristly wings down their inner side and a large hood over the mouth of the pot. They are often finely marked with deep blood-red streaks or even red all over. The upper pitchers are longer and taper elegantly from the top downwards into a curved base. The fluted toothed mouth to the pitcher is often red-striped.

Amongst the many paintings of Bornean plants in the archives of the Royal Botanic Gardens, Kew are ten coloured drawings of *Nepenthes* attributed to Sir James Brooke. Three plates illustrate the ground and aerial pitchers of *N. rafflesiana* which is commonly found in the vicinity of Kuching. It seems possible that they came to Kew at the same time as Hugh Low's drawings, being annotated by the same hand as the latter's drawings. It is possible that they were donated to Kew along with Low's drawings. They are of a uniformly high quality, rather better as botanical art than Low's more stylised portraits of the same species. Two copies exist of three of the drawings and lead one to suspect that they were drawn by a local artist for James Brooke, rather than by his own hand.

Alpinia glabra Ridl. (Plate 59)

Hugh Low drew two species of ginger, a family well represented by spectacular species in Borneo. *Alpinia glabra*, one of several species of the genus to be found on the island, is one of the species with a terminal inflorescence on leafy stems. Others have inflorescences on fertile stems separate from the leafy sterile ones. This species, one of several in the large tropical Asiatic genus *Alpinia*, was described in 1899 by Henry Ridley, several years after Low had drawn it. It grows to 1.5 m tall and is found in hill and montane forest.

Burbidgea schizocheila Hackett (Plate 60)

The orange flowers of *Burbidgea schizocheila*, a wild ginger, are frequently seen on the roadsides and forest margins of the mountains of north-west Borneo, including Mt Kinabalu. The genus was named by Sir Joseph Hooker in honour of Frederick William Burbidge (1847–1905), who collected plants in Australia, the Southwest Pacific islands and Borneo, especially on Mt Kinabalu, in 1877 for Messrs James Veitch & Sons. This species was described in 1904, many years after it had been drawn by Hugh Low.

Hoya coronaria Blume (Plate 61)

Hoyas of are common climbing vines throughout the forests of tropical Asia, Australia and the Pacific Islands. The leaves are leathery and the waxy flowers are borne in semi-globose to flat heads. Like many members of the family Asclepiadaceae, the stems and leaves weep white latex when damaged. *Hoya coronaria*, described by the great Dutch botanist Carl Blume in 1826, is one of the larger flowered species of the genus. It is found in hill and lowland forests up to about 900 m elevation. It is common on the lower slopes of Mt Kinabalu where Low probably encountered it on his expeditions to the mountain.

Costus speciosus (Koenig) J.E. Smith **(no ill.)**

Costus speciosus is a member of the family Costaceae, a close ally of the gingers (Zingiberaceae). It is widespread through the region from

Thailand southwards to the Malay Archipelago. It is a herb ranging from 1.5 to 3.5 m tall, the lower part of the stem is bare and the upper part leafy with the leaves borne in a spiral. The inflorescence is terminal, although some plants branch with each branch bearing a terminal inflorescence. The bracts are red or orange and contrast well with the flowers which are white with a yellow lip. The seeds are black. The young shoots are cooked and eaten as a vegetable in Borneo or are cooked and served in a soup.

It grows in secondary forest or on the edges of forest, often by streams and rivers in lowland Dipterocarp forest or in low open forest over limestone.

Dolichandrone spathacea K. Schumann (syn. *D. rheedii* Seem.) (Plate 62)

A widespread tree throughout the region, *Dolichandrone spathacea* is found growing in swamp forest or in secondary forest by rivers. It can reach about 15 m in height and has long tubular flowers that are 15–18 cm long. The paired fruits that hang down are black when ripe and up to 50 cm long. It is nearly always found on flat land, never in the hills.

Ixora pyrantha Bremekamp (Plate 66)

This fine ornamental species of *Ixora*, a member of the coffee family (Rubiaceae), is one of many similar species found in Borneo and the surrounding region. There are several scarlet-flowered species of *Ixora* in Borneo. One of these is the fine scarlet-flowered species collected by Hugh Low in Sarawak on January 17th 1845. He noted: 'I have made several interesting excursions in the jungle and have found ten interesting plants, viz. A willow-leaved *Ixora*…'.

Ixora pyrantha is a shrub up to 3.5 m tall that was described from a plant collected in Kuching by Mr Haviland. It can have red, pink or golden flowers. Its fruits are green but turn purple-black on ripening. It usually grows in lowland and mixed hill Dipterocarp forest. In Borneo, ixoras are used as a cure for headaches and tooth-ache.

Ploiarium alternifolium (Vahl) Melchior (syn. *Archytaea vahlii* Choisy)
(Plate 63)

Although previously considered a member of the Theaceae, *Ploiarium* is now placed in the family Guttiferae (Hypericaceae). *Ploiarium alternifolium* is found from Thailand and Cambodia south into the Malay Archipelago. It grows in swamp forest, in hill forest and kerangas in Borneo. It can be either a shrub or a tree up to 15 m tall, depending on its habitat. The leaves are characteristically bright green with a red stalk, veins and margin. The flowers are pinkish white or white with a red outer surface, the many stamens have white filaments and yellow anthers. Its wood is widely used for building purposes and for firewood.

REFERENCES

Desmond, R. (1992). *The European discovery of the Indian flora*. Oxford University Press.

ACKNOWLEDGEMENTS

I would like to thank the Keeper of the Herbarium, Royal Botanic Gardens, Kew for allowing the use of the facilities; Ms Marilyn Ward for assistance in locating the paintings; and Jeffrey Wood for help in identifying the subjects of the orchid portraits and Martin Cheek for the *Nepenthes*. Permission to use the illustrations is gratefully acknowledged to the Director and the Trustees of the Royal Botanic Gardens, Kew.

35 RENANTHERA LOWII

Nº 5 of Cut
Aug 28th 1879

Plate 1. *Dimorphorchis lowii* drawn by John Day.

Plate 2. *Paphiopedilum lowii* drawn by John Day.

Plate 3. *Dendrobium lowii* drawn by John Day.

Plate 4. *Plocoglottis lowii*

Plate 5. *Trichoglottis philippinensis* Lindl. (right),
Dendrobium crumenatum (left).

Plate 6. *Dendrobium tenue* (left), *Cleisostoma teretifolium* (right)

Plate 7. *Calanthe pulchra*

Plate 8. *Bulbophyllum brienianum* (Rolfe) J.J. Sm.

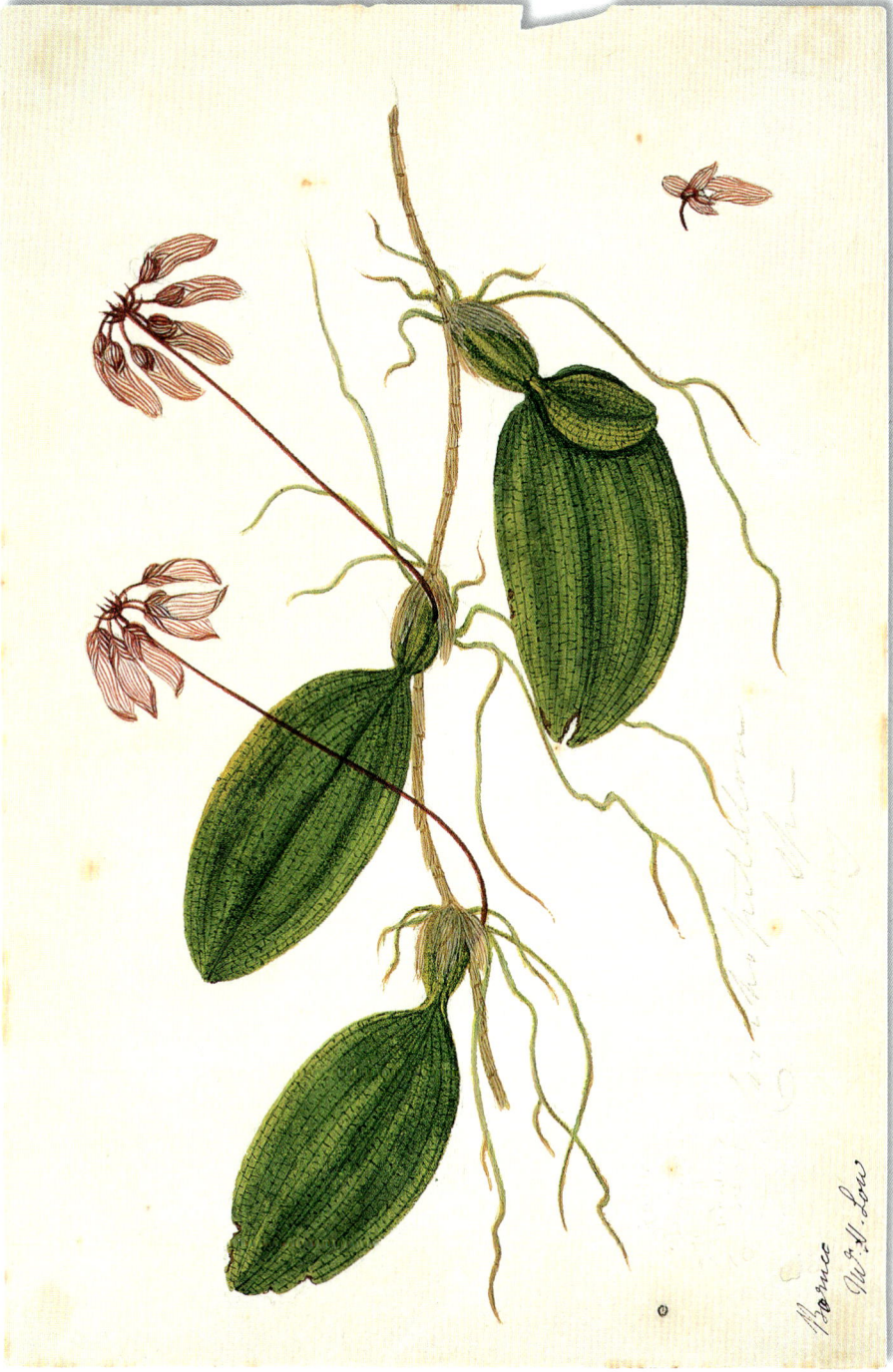

Plate 9. *Bulbophyllum lepidum* (Blume) J.J. Sm.

Plate 10. *Bulbophyllum purpurascens* Teijsm. & Binn.

Plate 11. *Bulbophyllum vaginatum* (Lindl.) Rchb.f.

Plate 12. *Dendrobium lamellatum* (Bl.) Lindl. (Right). *Thelasis* sp. (Left).

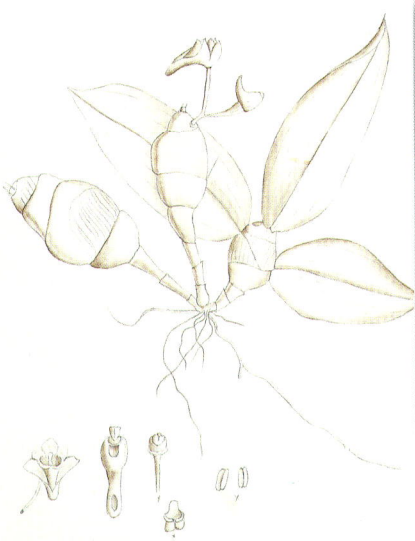

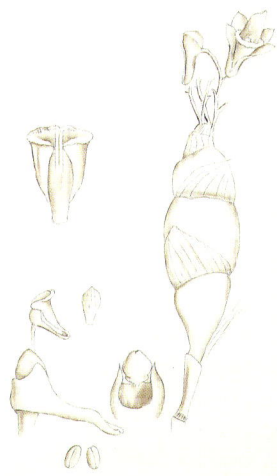

Plate 13. *Dendrobium lamellatum* (Bl.) Lindl.

Plate 14. *Dendrobium pachyanthum* Schltr.

Plate 15. *Dendrobium pachyanthum* Schltr.

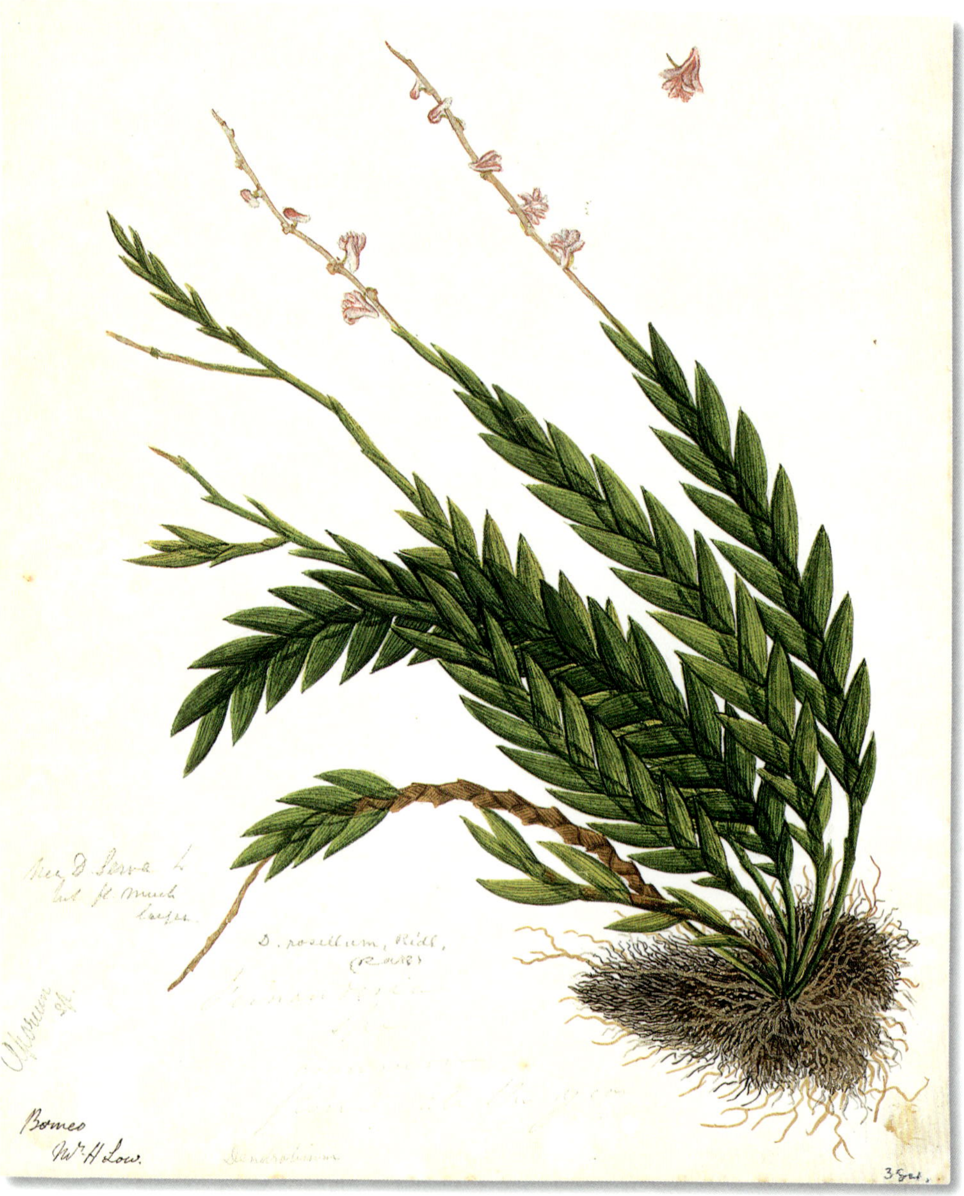

Plate 16. ***Dendrobium rosellum*** **Ridl.**

Plate 17. *Dendrobium spurium* (Bl.) J.J. Sm.

Borneo
W. H. Low.
Dendrobium criniferum, Lindl.
8 Cadetia.   (P. n. a.)

Plate 18. *Flickingeria comata* (Bl.) A.D. Hawkes

near E cinnabarina, Rolfe

See North Gallery
n. 554. (R. a. R.)

Borneo.
W. H. Low.

Plate 19. *Eria ignea* Rchb.f.

Plate 20. *Eria ignea* Rchb.f.

Plate 21. *Eria pulchella* Lindl.

Plate 22. *Eria pulchella* Lindl.

Plate 23. *Eria cepifolia* Ridl.

Plate 24. *Eria hyacinthoides* (Bl.) Lindl.

Plate 25. *Eria linearifolia* Ames

Borneo
W. H. Low

Eria vestita L.

Plate 26. *Trichotosia vestita* (Lindl.) Kraenzl.

Borneo.
W. H. Low.

Polycheilos cornu-cervi
(Phalaenopsis)
K.H.
(fide R.b.K)

Plate 27. *Phalaenopsis cornu-cervi* (Breda) Blume & Rchb.f.

Plate 28. *Luisia antennifera* Blume

Plate 29. *Micropera fuscolutea* (Lindl.) Garay

Borneo
Mr H. Low.

Vanda

Acampe?

Plate 30. *Trichoglottis geminata* J.J. Sm.

Plate 31. *Trichoglottis tinekeae* Schmit. (Left). *Thrixspermum* sp. (Right).

Borneo
W. H. Low.

Sarcochilus
Angraecum fragmultos

Plate 32. *Thrixspermum centipeda* Lour.

Plate 33. *Thrixspermum centipeda* Lour.

Borneo
W. H. Low
Appendicula ?

Plate 34. *Appendicula* sp.

Plate 35. *Annona muricata* L.

Plate 36. *Annona squamosa* L.

Plate 37. *Garcinia mangostana* L.

Plate 38. *Nephelium lappaceum* L.

Plate 39. Syzygium aqueum (Burm.f.) Alston (as *Eugenia*)

Plate 40. *Musa paradisiaca* L. (syn. *M. sapientium*)

Plate 41. *Ananas comosus* (L.) Merr.

Plate 42. *Artocarpus heterophyllus* Lamark

Plate 43. *Tamarindus indica* L.

Plate 44. *Theobroma cacao* L.

Plate 45. *Coffea arabica* L.

Plate 46. *Piper nigrum* L.

Plate 47. *Myristica fragrans* Houtt.

Plate 48. *Lepeostegeres beccarii* (King) Gamble

Plate 49. *Lepidaria sabaensis* (Stapf) Tiegh.

N ampullaca.

Borneo
Mr H. Low.

Plate 50. *Nepenthes ampullaria* Jack

N. ampullaria

Borneo

Sir J. Brooke

Plate 51. *Nepenthes ampullaria* Jack drawn by James Brooke

Plate 52. *Nepenthes gracilis* Korthals

Borneo
Mr H. Low.

N. tentaculata.

Plate 53. *Nepenthes gracilis* Korthals

Plate 54. *Nepenthes gracilis* Korthals

Plate 55. *Nepenthes* ×*trichocarpa*

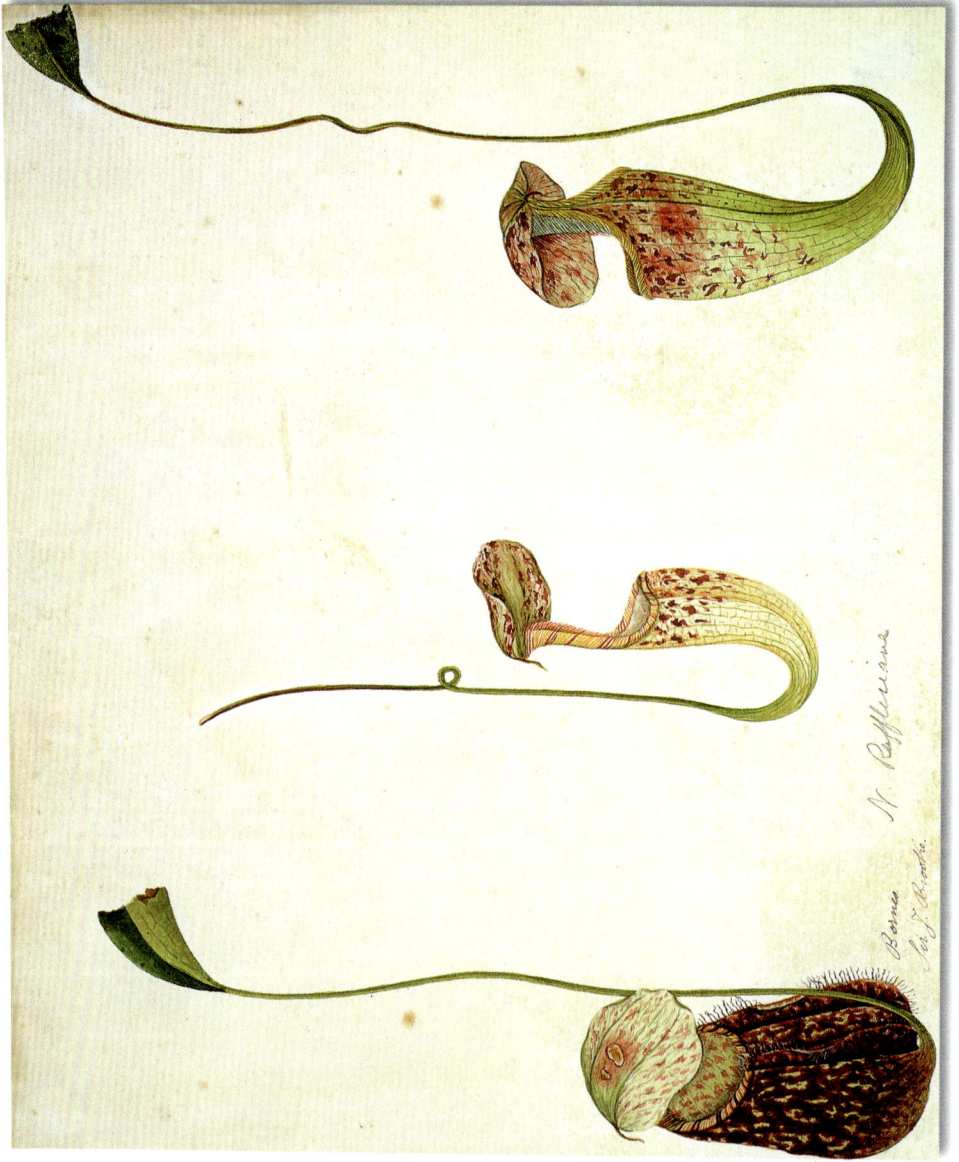

Plate 56. *Nepenthes rafflesiana* Jack drawn by James Brooke

N. Rafflesiana

Borneo
Sir J. Brooke. 136.

Plate 57. *Nepenthes rafflesiana* Jack drawn by James Brooke

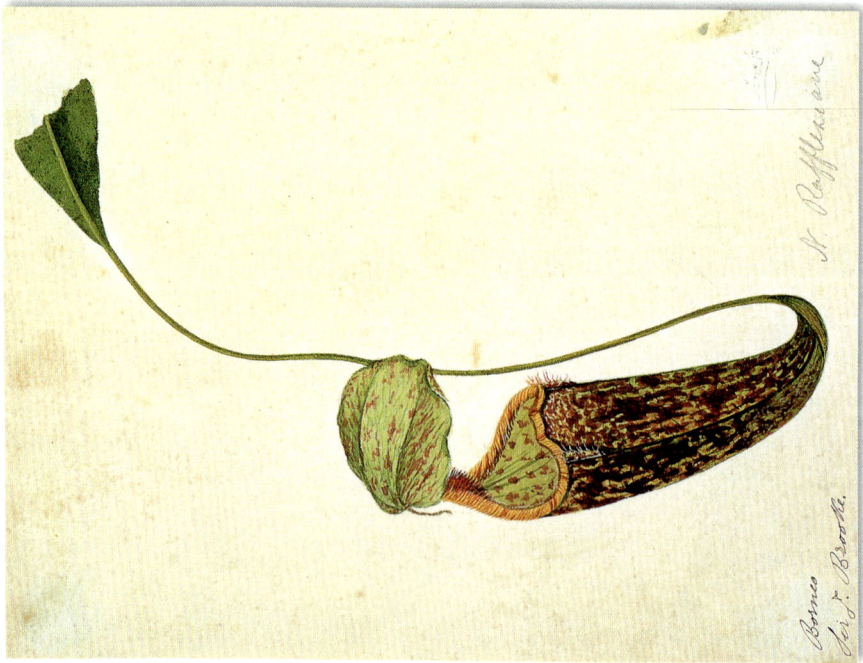

Plate 58. *Nepenthes rafflesiana* Jack drawn by James Brooke

Plate 59. *Alpinia glabra* Ridl.

Borneo
Mr. H. Low.

f. Burbidgea

Plate 60. *Burbidgea schizocheila* Hackett

Plate 61. *Hoya coronaria* Blume

Plate 62. *Dolichandrone spathacea* K. Schumann
(syn. *D. rheedii* Seem.)

Plate 63. *Ploiarium alternifolium* (Vahl) Melchior
(syn. *Archytaea vahlii* Choisy)

Plate 64. *Artocarpus altilis* (Parkinson) Fosberg

Plate 65. *Syzygium aromaticum* (L.) Merrill & Perry

Borneo.

H. Low Esqr

Ixora
not coccinea

Plate 66. *Ixora pyrantha* Bremekamp

INDEX OF PLANT NAMES

Compiled by Phillip Cribb and Cheng Jen Wai

Titles by Natural History Publications (Borneo)

For more information, please contact us at

Natural History Publications (Borneo) Sdn. Bhd.
A913, 9th Floor, Phase 1, Wisma Merdeka
P.O. Box 15566, 88864 Kota Kinabalu, Sabah, Malaysia
Tel: 088-233098 Fax: 088-240768 e-mail: chewlun@tm.net.my
www.nhpborneo.com

Head Hunting and the Magang Ceremony in Sabah by Peter R. Phelan

A Botanist in Borneo: Hugh Low's Sarawak Journals, 1844–1846 (Edited and introduced by R.H.W. Reece and with notes on Hugh Low's plant portraits by P.J. Cribb)

Mount Kinabalu: Borneo's Magic Mountain—an introduction to the natural history of one of the world's great natural monuments by K.M. Wong and C.L. Chan

On the Flora of Mount Kinabalu in North Borneo by Otto Stapf. Reprint with an Introduction by John H. Beaman

A Contribution to the Flora and Plant Formations of Mount Kinabalu and the Highlands of British North Borneo by Lilian S. Gibbs. Reprint with an Introduction by John H. Beaman

Discovering Sabah by Wendy Hutton (English, Chinese and Japanese editions)

Enchanted Gardens of Kinabalu: A Borneo Diary by Susan M. Phillipps

A Colour Guide to Kinabalu Park by Susan K. Jacobson

Kinabalu: The Haunted Mountain of Borneo by C.M. Enriquez (Reprint)

National Parks of Sarawak by Hans P. Hazebroek and Abang Kashim Abg. Morshidi

A Walk through the Lowland Rainforest of Sabah by Elaine J.F. Campbell

In Brunei Forests: An Introduction to the Plant Life of Brunei Darussalam (Revised edition) by K.M. Wong

The Larger Fungi of Borneo by David N. Pegler

Rafflesia of the World by Jamili Nais

Pitcher-plants of Borneo by Anthea Phillipps and Anthony Lamb

A Field Guide to the Pitcher Plants of Sabah by Charles Clarke

Nepenthes of Borneo by Charles Clarke

Nepenthes of Sumatra and Peninsular Malaysia by Charles Clarke

The Plants of Mount Kinabalu 3: Gymnosperms and Non-orchid Monocotyledons
  by John H. Beaman and Reed S. Beaman

The Plants of Mount Kinabalu 4: Dicotyledon Families Acanthaceae to Lythraceae by
  John H. Beaman, Christiane Anderson and Reed S. Beaman

Slipper Orchids of Borneo by Phillip Cribb

The Genus Paphiopedilum (Second edition) by Phillip Cribb

Orchids of Sarawak
  by Teofila E. Beaman, Jeffrey J. Wood, Reed S. Beaman and John H. Beaman

Orchids of Sumatra by J.B. Comber

Dendrochilum of Borneo by J.J. Wood

The Genus Coelogyne: A Synopsis by Dudley Clayton

Gingers of Peninsular Malaysia and Singapore
  by K. Larsen, H. Ibrahim, S.H. Khaw and L.G. Saw

Mosses and Liverworts of Mount Kinabalu
  by Jan P. Frahm, Wolfgang Frey, Harald Kürschner and Mario Manzel

Birds of Mount Kinabalu, Borneo by Geoffrey W.H. Davison

The Birds of Borneo (Fourth edition)
  by Bertram E. Smythies (Revised by Geoffrey W.H. Davison)

The Birds of Burma (Fourth edition)
  by Bertram E. Smythies (Revised by Bertram E. Smythies)

Swiftlets of Borneo: Builders of Edible Nests
  by Lim Chan Koon and Earl of Cranbrook

Proboscis Monkeys of Borneo by Elizabeth L. Bennett and Francis Gombek

The Natural History of Orang-utan by Elizabeth L. Bennett

A Field Guide to the Frogs of Borneo by Robert F. Inger and Robert B. Stuebing

A Field Guide to the Snakes of Borneo by Robert B. Stuebing and Robert F. Inger

Man-eating Crocodiles of Borneo by James Ritchie with Johnson Jong

Turtles of Borneo and Peninsular Malaysia by Lim Boo Liat and Indraneil Das

The Natural History of Amphibians and Reptiles in Sabah
  by Robert F. Inger and Tan Fui Lian

An Introduction to the Amphibians and Reptiles of Tropical Asia by Indraneil Das

Marine Food Fishes and Fisheries of Sabah by Chin Phui Kong

Layang Layang: A Drop in the Ocean
  by Nicolas Pilcher, Steve Oakley and Ghazally Ismail

Phasmids of Borneo by Philip E. Bragg

The Dragon of Kinabalu and other Borneo Stories by Owen Rutter (Reprint)

Land Below the Wind by Agnes N. Keith (Reprint)

Three Came Home by Agnes N. Keith (Reprint)

White Man Returns by Agnes N. Keith (Reprint)

Forest Life and Adventures in the Malay Archipelago by Eric Mjöberg (Reprint)

A Naturalist in Borneo by Robert W.C. Shelford (Reprint)

Twenty Years in Borneo by Charles Bruce (Reprint)

With the Wild Men of Borneo by Elizabeth Mershon (Reprint)

Kadazan Folklore (Compiled and edited by Rita Lasimbang)

A Cultural Heritage of North Borneo: Animal Tales by P.S. Shim

An Introduction to the Traditional Costumes of Sabah
  (eds. Rita Lasimbang and Stella Moo-Tan)

Bahasa Malaysia titles:

Manual latihan pemuliharaan dan penyelidikan hidupan liar di lapangan
  oleh Alan Rabinowitz (Translated by Maryati Mohamed)

Etnobotani oleh Gary J. Martin (Translated by Maryati Mohamed)

Panduan Lapangan Katak-Katak Borneo oleh R.F. Inger dan R.B. Stuebing

Other titles available through
Natural History Publications (Borneo)

The Bamboos of Sabah by Soejatmi Dransfield

The Morphology, Anatomy, Biology and Classification of Peninsular Malaysian Bamboos by K.M. Wong

Orchids of Borneo Vol. 1 by C.L. Chan, A. Lamb, P.S. Shim and J.J. Wood

Orchids of Borneo Vol. 2 by Jaap J. Vermeulen

Orchids of Borneo Vol. 3 by Jeffrey J. Wood

Orchids of Java by J.B. Comber

Forests and Trees of Brunei Darussalam (Edited by K.M. Wong and A.S. Kamariah)

A Field Guide to the Mammals of Borneo by Junaidi Payne and Charles M. Francis

Pocket Guide to the Birds of Borneo Compiled by Charles M. Francis

Kinabalu: Summit of Borneo (Edited by K.M. Wong and A. Phillipps)

Ants of Sabah by Arthur Y.C. Chung

Traditional Stone and Wood Monuments of Sabah by Peter Phelan

Borneo: The Stealer of Hearts by Oscar Cooke (Reprint)

Maliau Basin Scientific Expedition (Edited by Maryati Mohamed, Waidi Sinun, Ann Anton, Mohd. Noh Dalimin and Abdul-Hamid Ahmad)

Tabin Scientific Expedition (Edited by Maryati Mohamed, Mahedi Andau, Mohd. Nor Dalimin and Titol Peter Malim)

Klias-Binsulok Scientific Expedition (Edited by Maryati Mohamed, Mashitah Yusoff and Sining Unchi)

Traditional Cuisines of Sabah (Edited by Rita Lasimbang)

Cultures, Costumes and Traditions of Sabah, Malaysia: An Introduction

Tamparuli Tamu: A Sabah Market by Tina Rimmer